Signals and Communication Technology

This series is devoted to fundamentals and applications of modern methods of signal processing and cutting-edge communication technologies. The main topics are information and signal theory, acoustical signal processing, image processing and multimedia systems, mobile and wireless communications, and computer and communication networks. Volumes in the series address researchers in academia and industrial R&D departments. The series is application-oriented. The level of presentation of each individual volume, however, depends on the subject and can range from practical to scientific.

Indexing: All books in "Signals and Communication Technology" are indexed by Scopus and zbMATH

For general information about this book series, comments or suggestions, please contact Mary James at mary.james@springer.com or Ramesh Nath Premnath at ramesh.premnath@springer.com.

Mourad Abbas
Editor

Mastering Text Classification

Cutting-Edge NLP Techniques

Springer

Editor
Mourad Abbas
University of Science and Technology
Houari Boumediene
Algiers, Algeria

ISSN 1860-4862 ISSN 1860-4870 (electronic)
Signals and Communication Technology
ISBN 978-3-031-93611-1 ISBN 978-3-031-93612-8 (eBook)
https://doi.org/10.1007/978-3-031-93612-8

This Springer imprint is published by the registered company Springer Nature Switzerland AG
The registered company address is: Gewerbestrasse 11, 6330 Cham, Switzerland

Preface

Text classification has evolved dramatically in recent years. In fact, it has undergone a profound transformation driven by the rise of deep learning and, more specifically, the pre-trained, Transformer-based language models like BERT, GPT, and their many variants leading to handle better human language inherent complexities.

Besides model architecture, human-centric approaches have an important impact on these models performance. For decades, text classification has been driven by the pursuit of a single *ground truth* label, a simplifying assumption that often fails in the face of real-world complexities. One key area is moving beyond this limiting assumption by incorporating the diverse perspectives of multiple human annotators. Disagreement between annotators in text classification is not a bug to be fixed but a feature to be leveraged. Relying on this principle, the book's opening chapter explores the idea of embracing this feature during training and prediction. We illustrate how a probability-based multi-label method, ensemble systems, and instruction fine-tuning can surpass conventional models by effectively capturing the rich, intricate, and multifaceted nature of human judgment on delicate topics, such as hate speech and abusive conversations.

The second chapter introduces how the unified dimension approach can be used for decomposing and classifying discourse relations across different frameworks, such as Rhetorical Structure Theory, Penn Discourse Treebank, and Segmented Discourse Representation Theory, highlighting how a deeper understanding of textual structure can enhance classification.

Building on the importance of linguistic features, we devoted two chapters to how Part-of-Speech tags can enhance Named Entity Recognition for Arabic and its dialects, challenging the specificity of this language and the lack of dialectal resources. The book broadens its scope to address multitask learning framework for multilingual text classification, aiming to improve accuracy and robustness, with a focus on low-resource languages, testing the model's ability to generalize to new languages from a different family. Following this, the next chapter explores new methods to address the limitations of existing deep hierarchical text classification techniques. The effectiveness of this approach is demonstrated by applying it to different domains like classifying medical procedures in Russian and organizing product titles in e-commerce.

It is my sincere hope that this book provides a new perspective to readers and serves as a useful resource for students and researchers alike. With these thoughts in mind, I invite you to explore the ideas and methods within these pages.

Algiers, Algeria Mourad Abbas

Contents

Contents

Embracing Disagreement in Text Classification During Training and Prediction

Jin Xu, Mariët Theune, and Daniel Braun

1 Introduction

It is a widely adopted practice that data is annotated by multiple annotators and the annotations are afterward aggregated, e.g., through majority voting, in order to mitigate biases and allow for error detection and correction [1]. However, genuine disagreement between annotators, which can provide valuable insights, e.g., for subjective tasks like detection of hate speech, emotions, or sexism, is also removed in the process [2, 3]. In recent years, the practice of only considering majority annotations has been increasingly criticized [4, 5]. Many researchers have started to advocate for better ways to deal with disagreement between annotators, shifting away from the conventional approach of adopting monolithic, majority-aggregated gold standards [6–9].

In our previous work [10], we proposed three different strategies to leverage annotator disagreement during the training of text classification models: a probability-based multi-label approach, an ensemble system approach, and an instruction tuning approach. We compared these strategies against a baseline model that is trained on

J. Xu · M. Theune
Faculty of Electrical Engineering, Mathematics, and Computer Science, University of Twente, Enschede, Netherlands
e-mail: jinxu130817@gmail.com; m.theune@utwente.nl

D. Braun (✉)
Department of Mathematics and Computer Science, University of Marburg, Marburg, Germany
e-mail: daniel.braun@uni-marburg.de

M. Abbas (ed.), *Mastering Text Classification*, Signals and Communication Technology, https://doi.org/10.1007/978-3-031-93612-8_1

the majority labels derived from the multiple annotations. In this chapter, in extension of our previous work, we explore the reasons why the proposed models outperformed the baseline model. We choose two text classification tasks which inherently entail a high degree of subjectivity for the evaluation: hate speech detection and abuse detection in conversations. In our chosen datasets [11, 12], these two tasks exhibit different complexity and difficulty in terms of the label space: While the hate speech detection dataset contains binary labels, the abusive conversation detection dataset is not only annotated with abusive/non-abusive but also the severity of the abuse.

Our first approach tackles the tasks as a probability-based multi-label text classification problem. Instead of predicting specific labels to one instance, the model provides a probability distribution. Each value in this distribution represents the likelihood regarding the instance's association with each label. The second approach imitates the process of annotation from multiple annotators with an ensemble system. The ensemble system consists of many sub-models, each of which is trained on different labels to capture the diverse viewpoints embedded in the annotations. Thirdly, we use instruction tuning. Specifically, we use a pretrained generative model and inject explicit guidance into the training process to customize the model's behavior. The performance of the proposed models is compared using cross entropy. Following this, we aim to evaluate the effectiveness of incorporating multiple labels against the model that only relies on the majority label. Due to the format disparity between the outputs generated by these two models, it is impractical to use cross entropy for evaluation. To bridge this gap, we conduct an online survey where participants specify their preference between probabilistic annotations generated from the probability-based multi-label model and the baseline model. We choose these two models because they share an identical structure, and their only difference is the labels they are trained on. In the survey, we also investigate the individuals' preference regarding the labeling method. Specifically, we aim to explore if individuals favor using a majority label or multiple labels to annotate the data.

The results show that on the hate speech dataset, the multi-label method outperforms the ensemble system and instruction tuning. Conversely, instruction tuning is the best-performing method on the abusive conversation dataset. Additionally, the multi-label model exhibits higher performance on the hate speech dataset compared to the abusive conversation dataset. By contrast, both the ensemble system and instruction tuning achieve better performance on the abusive conversation dataset. For the online survey, the results show that the outputs from the multi-label model are considered more reasonable than those from the baseline model to characterize samples from the online survey. This proves the effectiveness of leveraging annotation disagreements for model training. The survey results also demonstrate that individuals do not show a notable preference of labeling methods between majority label or multiple labels. However, a qualitative analysis of the results reveals that the labeling method preference appears to be influenced by how emotions or opinions are expressed in the text. For the hate speech dataset, individuals' preferences for probability annotations can be influenced by their chosen labeling method. In the abusive conversation dataset, however, individuals tend to prefer softer distributions over harsher ones as representations of abusive conversations.

The key contributions of this chapter can be summarized as follows:

- We construct models that can accommodate diverse perspectives by considering the individual labels from multiple annotators via three different approaches.
- We investigate the performance of these proposed models across the two datasets, which differ in aspects such as data size, the number of annotators involved in each sample, classification difficulty, etc.
- We carry out an online survey to evaluate the effectiveness of incorporating multiple labels for model training against the model that only considers the majority label. This survey facilitates a comparison between these two models by addressing the differences in their output formats.

The remainder of this text is structured as follows: Sect. 2 presents an overview of related work. Section 3 discusses our two experimental datasets. The methodology is illustrated in Sect. 4. Subsequently, Sect. 5 gives a description of the experimental results for both cross entropy and human preference, and a comprehensive analysis. Finally, the limitations and conclusions are summarized.

2 Related Work

Frenda et al. [13] recently presented a comprehensive survey of literature on perspectivist approaches in NLP. Uma et al. [7] presented a broader overview of literature on learning from disagreement that also includes work from computer vision, in 2021. Here, we will focus on related work regarding the sources of disagreement and NLP methods that leverage disagreement.

2.1 Sources of Disagreement

Disagreement in annotations can originate from different sources, such as the inherent ambiguity of text subjectivities and variations in value systems of annotators. Natural language, especially texts, can be inherently complex and interpreted in multiple ways within a given context [14]. There are many subjective elements which may add an additional layer of intricacy to the understanding of texts, such as sentiments, opinions, or nuanced expressions. Therefore, it is common that there are divergent interpretations among annotators. Furthermore, some sentences and even the definition of labels may contain vague or ambiguous statements [15], making it challenging for annotators to reach an agreement.

However, annotators themselves and their background can also have significant impact on the annotation results, such as cultural differences, individual value systems or personal discrepancies, etc. [16]. Through post-annotation interviews, Patton et al., for example, showed that annotators who come from communities discussed in gang-related tweets are more likely to rely on their lived experiences in the process of annotating when compared to graduate student researchers [17]. This divergence results in distinct label judgments. Additionally, Luo et al. found that the political affiliation of annotators can significantly shape how they assess and annotate political stances [18].

2.2 Methods Leveraging Disagreement

Majority voting involves aggregating annotations by selecting the label that the majority of annotators agree upon. It is easy to understand and implement and tends to perform well when the annotators share unanimous perspectives [7]. However, the employment of a majority voting method in annotation processes can unintentionally obscure nuanced viewpoints, especially for groups that are underrepresented in annotator pools [19]. To address this concern, it is important to ensure a diverse representation among annotators to foster a more comprehensive understanding of various perspectives, particularly those from underrepresented demographics [20]. Recognizing this necessity is crucial for fostering inclusivity and preventing the marginalization of specific viewpoints in annotation tasks. While majority voting is the most common approach, other aggregation methods exist to derive a "gold standard" from multiple, possibly disagreeing, annotations, like the use of an (more senior) expert annotator that is used to decide in case of disagreements [9].

Some studies have introduced alternative methods to label aggregation in order to incorporate annotator disagreement in model training. For example, Chou et al. incorporated the characteristics of each annotator in the inner layers of the neural network [21]. In their experiment, they also introduced a joint learning methodology that modeled the label uncertainty and annotator idiosyncrasy simultaneously by using both hard label (majority voting) and soft label (the distribution of annotations). The results showed that the soft label contains useful information that significantly boosts the model performance. Fornaciari et al. proposed a multitask neural network that was trained on soft label distribution over annotator labels [22]. By integrating a divergence measurement between soft label and "true" label vectors into the loss functions, they effectively mitigated overfitting and therefore improved model performance. Davani et al. introduced multi-annotator models where each annotator's judgments were regarded as an independent subtask with a shared common representation of the annotation task [16]. This approach enables to preserve and model the internal consistency in each annotator's labels. It also incorporates the systematic disagreements with other annotators. Similarly, the network architecture introduced by Guan et al. individually models annotation experts [23]. In this approach, each expert's model weight is calculated independently, and these individual weights are then averaged to facilitate ensemble recognition. To include the knowledge from annotators, Fayek et al. employed neural networks to build an ensemble system that consists of many models, with each model representing one annotator [24]. Then the final results are obtained by combining the individual model outputs. For the purpose of comparison, they also trained a model that is fed with the soft labels from all annotators. The results showed that these two approaches exhibit similar performance, which means that the performance improvement from the ensemble could be achieved by using soft labels from annotators.

Although the approaches outlined above have improved the performance by leveraging annotation disagreements, they remained limited to identifying the majority label. The outputs, in the form of "soft labels" (probability distribution over labels), were still aggregated to single labels as final predictions. Accordingly, there is limited research focusing on evaluating the effectiveness of embracing multiple labels. To address the first gap, this study proposes three approaches for constructing text classification models that output "soft labels." The handling of the second gap involves conducting an online survey, which will be discussed in Sect. 4.5.

3 Datasets

3.1 Hate Speech

The first dataset is the "Large-Scale Hate Speech Dataset"[1] (hereinafter referred to as "hate speech dataset") introduced in [11]. It consists of a total of 100,000 tweets (70,000 training, 15,000 validation, and 15,000 testing). Each tweet is annotated by five annotators that have been selected randomly from a panel of 20 annotators. According to the annotation guidelines utilized by Sharma et al. [25], tweets are categorized as "Hate" if they target, incite violence against, threaten, or advocate for physical harm toward an individual or a group of people based on identifiable trait or characteristic. If tweets humiliate, taunt, discriminate against, or insult an individual or a group of people, they are annotated as "Offensive." In the absence of these criteria, the tweets are labeled as "Normal." An example of a data sample is demonstrated in Table 1. The "HateLabel" is determined by majority voting across all five annotations. In cases of a tie situation, a dataset curator outside the initial annotator set determines the label as the final decision.

3.2 Abusive Conversation

The second dataset is the "Abuse in Conversational AI" dataset[2] (hereinafter referred to as "abusive conversation dataset") introduced in [12]. It was collected from conversations between users and conversational AI systems and consists of 2501 samples as training data, 831 as validation data, and 853 as testing data. The data was annotated

[1] https://github.com/avaapm/hatespeech/tree/master/dataset_v1

[2] https://github.com/amandacurry/convabuse/tree/main

Table 1 Excerpt from the hate speech dataset [11]

Column name	Description	Example
Text	Tweet's text contents	I don't care what this man's beliefs were, no one deserves to die this way. Hopefully this event will help raise awareness......
Label_1	Annotation of the first annotator	Offensive
Label_2	Annotation of the second annotator	Normal
......	......	......
Label_5	Annotation of the fifth annotator	Hate
HateLabel	Majority label decision	Normal

Table 2 Excerpt from the abusive conversation dataset [12]

Column name	Description	Example
prev_agent	Agent's previous utterance	How long have you been on the verge of suicide?
prev_user	User's previous utterance	Since I was 10 years old
Agent	Agent's utterance	Why do you tell me you were 10 years old just now?
User	User's utterance	Were you not 10 years old?
Annotator1.1	Not abusive	1
Annotator1.0	Ambiguous	0
Annotator1.-1	Mildly abusive	0
Annotator1.-2	Strongly abusive	0
Annotator1.-3	Very Strongly abusive	0
......	......	......
Annotator8.-3	Very Strongly abusive	/

using an unbalanced rating scale proposed by Poletto et al. [26], ranging from *not abusive*, *ambiguous*, *mildly abusive*, *strongly abusive* to *very strongly abusive*. This annotation scheme offers insights into not only the presence of abusive content, but also the severity of the abuse. Specifically, *very strongly abusive* denotes content that is strongly negative with overt incitement to hatred, violence, or discrimination and an attitude geared toward attacking or demeaning the target. *Strongly abusive* indicates content that is negative, insulting, or abusive, with an aggressive tone. *Mildly abusive* represents content that is negative and impolite, mildly offensive but still within the bounds of conversational language. *Ambiguous* suggests ambiguity or unclear sentiment, while a *not abusive* indicates the non-abusive content. In the annotation process, eight annotators were recruited, and each example is annotated by a minimum of three annotators. Table 2 gives an example from this dataset. One sample consists of two rounds of dialogue between the user and agent. For annotators not assigned to a particular sample, their entries remain blank.

4 Methodology

This section provides details on the baseline model, the three proposed models, and the evaluation methods. The training hyperparameters for the different models can be found in Tables 8 and 9 in Appendix 1.

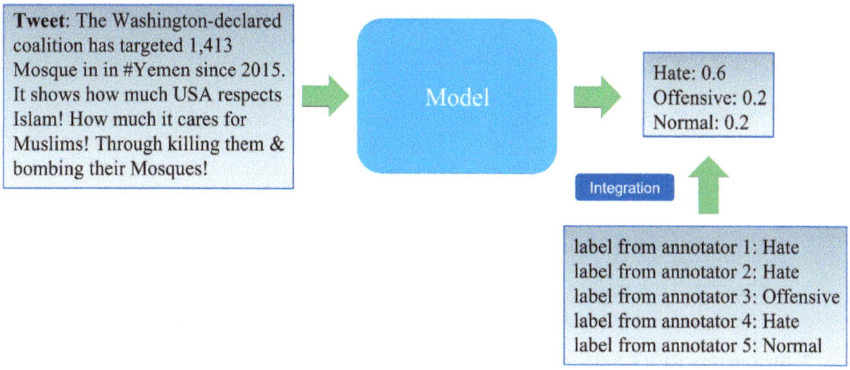

Fig. 1 The framework of model training within the probability-based multi-label method

4.1 Baseline Model

The baseline model is trained on the "ground truth" label that is aggregated via majority voting. Given BERT's [27] notable performance in contextual understanding, we choose it as the pretrained model. We augment its architecture by adding a fully connected layer to the last hidden state, thereby adapting the model structure to the specific prediction task.

4.2 Probability-Based Multi-label Method

The task of identifying hate speech or abusive conversation can be regarded as a multi-label text classification problem, where a given piece of text can be associated with one or multiple labels simultaneously. This paradigm is relevant in scenarios where content demonstrates diverse characteristics, and can be associated with various topics, themes, or attributes. Unlike the traditional approaches that assign one or several exclusive labels to the input text [28], our model predicts the probability of each label being associated with the given text. The approach is illustrated in Fig. 1. The model is trained on the probability distribution across different labels which is derived from individual annotations. Like the baseline model, the multi-label model is also based on BERT but fine-tuned with different types of target labels.

4.3 Ensemble System

In the annotation process, multiple labels are assigned by different annotators. Inspired by this and a similar idea from Fayek et al. [24], we propose an ensemble system consisting of several sub-models to simulate the process of annotation by

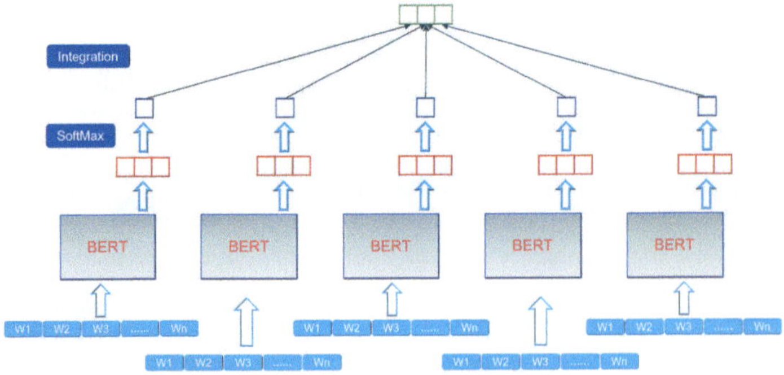

Fig. 2 Fine-tuning BERT individually as sub-models within the ensemble system

leveraging the diversity in labels provided by different annotators. The foundation model used to train on individual annotations is the pretrained BERT model. As shown in Fig. 2, each sub-model is based on a BERT model that is fine-tuned individually on its respective set of labels. For each sub-model, the input is the text from one sample, and the output is a multidimensional vector where each dimension corresponds to one category. After that, this vector is transformed by the SoftMax function, and the dimension with the highest probability is identified as the output of the sub-model. Finally, the predictions from all sub-models are combined and converted into a probability distribution with three or five dimensions. Within the ensemble system, each sub-model contributes a unique perspective to the final predictions, collectively representing the diversity and nuances present in the annotations. In this way, it aims to achieve a more comprehensive understanding of the data by incorporating the varied viewpoints of multiple annotators.

In the abusive conversation dataset, the annotators assigned for each sample are clearly specified and identifiable, as detailed in Table 2. Therefore, within the ensemble system, each sub-model represents one specific annotator and is trained on that annotator's labels. By contrast, in the hate speech dataset, each sample is labeled by five anonymous annotators. Despite the anonymity, training a model with such labels can potentially increase the robustness of sub-models since it helps to reduce the biases or inconsistencies introduced by individual annotators [29]. Furthermore, the resulting labels are likely to reflect a diverse range of perspectives and interpretations of the data. Training sub-models on these diverse annotations can capture the variability in annotator judgments and enhance the model's ability to generalize across different viewpoints [30]. This collective wisdom pooled from multiple annotators has the potential to improve overall performance.

Since the sub-models can show varying performances in the training and validation processes, the top n ($n \geq 3$) best-performing sub-models are chosen to determine the final output. The ranking is based on their accuracy on the validation data.

4.4 Instruction Tuning

Instruction tuning is the process of fine-tuning LLMs in a supervised fashion on a dataset consisting of pairs of instructions and outputs. Here, an "instruction" is constructed to guide the learning process. The key idea is to provide the model with explicit instructions to enhance its performance and align it with specific objectives. Unlike traditional training approaches where models learn from data alone, instruction tuning injects explicit guidance into the training process. This approach allows for explicit customization of the model's behavior. In this chapter, we ask the model to predict the class of hate speech or abusive conversation based on the input we construct. The input contains the task description, the instruction, the original text, and the annotation from one specific annotator (i.e., *not* the majority label). A template is needed to transform (text, label) pairs to (instruction, output) pairs by filling templates with data instances from the dataset [31]. Here, we utilize LLaMa2 (7 billion version)[3] as the foundation model. The fine-tuning process conducted via instruction tuning on the two datasets is presented in Figs. 3 and 4. On the left sides of the figures are the inputs fed into the pretrained model. Through fine-tuning, the model gains the ability to predict the corresponding label assigned by a particular annotator. In the inference phase, the value under the "Response" key will be removed, and the fine-tuned model is tasked with generating its prediction.

Due to the large parameter size of the pretrained model, fine-tuning the entire model can be time intensive and resource demanding. To address this, LoRA (low-rank adaptation) [32] is employed. With LoRA, the pretrained weights are frozen, and an adaptor is added alongside the base model. The adaptor is decomposed into two smaller matrices, which have much fewer parameters than the base model. During fine-tuning, only the parameters in the adaptor are updated, reducing the computations dramatically. During inference, the two matrices are multiplied and merged with base model, which ensures that the parameter size of newly fine-tuned model will stay unchanged. Like the ensemble system, the top n ($n \geq 3$) best-performing sub-models are selected to contribute to the final predictions.

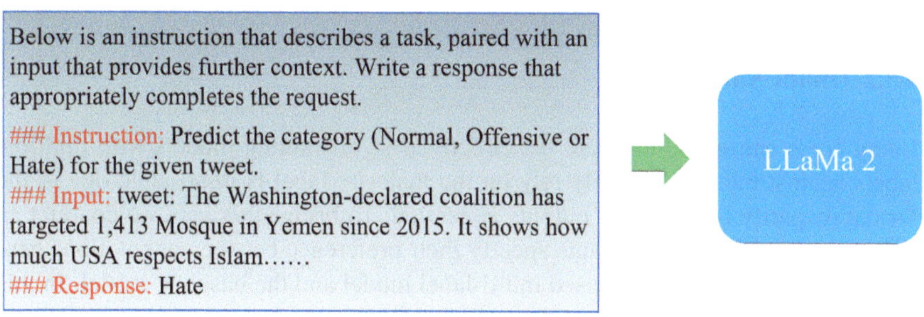

Fig. 3 Fine-tuning LLaMa 2 as a sub-model with instruction tuning in the hate speech dataset

[3] https://huggingface.co/meta-llama/Llama-2-7b-chat-hf

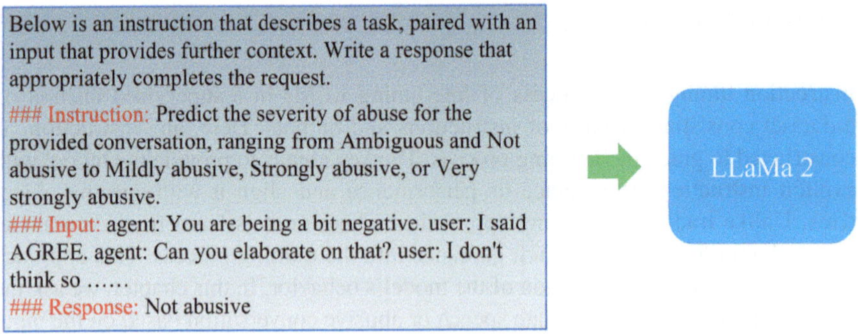

Fig. 4 Fine-tuning LLaMa 2 as a sub-model with instruction tuning in the abusive conversation dataset

4.5 Evaluation

4.5.1 Cross Entropy

The final output of our proposed models is a probability distribution across different labels, and in this scenario, a single "ground truth" label is no longer applicable for model evaluation. Instead, we use cross entropy to compare the distribution of annotations with model output. As a statistical distance, cross entropy measures how a probability distribution is different from a reference probability distribution. For two discrete probability distributions A and B which have the same class space χ, the way of calculating cross entropy between A and B is shown in formula (1) [33]. This metric is commonly used as the loss function for training neural networks. In the field of NLP, it has also been used to quantify how well the model's predicted distribution matches the annotation distribution [34].

$$H(A, B) = -\sum_{x \in \chi} a(x) \log b(x). \tag{1}$$

4.5.2 Online Survey

Using cross entropy to evaluate the effectiveness of training models with multiple labels against models that only rely on the majority label is impossible due to the format disparity between the outputs generated. To bridge this gap, we conduct an online survey where participants specify their preference between annotations generated from the probability-based multi-label model and the baseline model. In this survey, we also aim to understand whether individuals favor using a majority label or multiple labels as a labeling method. For each dataset, we select 10 samples, each featuring two groups of annotations. In the first group, there are two annotations. One is the multiple labels from all the annotators, and the other one is the "ground truth" label derived from these multiple labels via majority voting. There are also two annotations in the second group. Both annotations are in the form of probability distributions across different labels. One is generated from the baseline model trained with the majority labels, which are, however, used to generate a probability distribution in the phase of inference. The other one is from the probability-based multi-label model. This model has the same structure as the baseline model, and their only difference is the labels they were trained on. For an example of the online survey, please refer to Appendix 2. We recruited a total of 36 participants, all of whom were students at the University of Twente. Detailed demographic information for these participants is provided in Appendix 3. They were required to indicate, for each group, which annotation they found more reasonable to characterize the tweet or the abusive conversation.

Regarding the sampling strategy for the online survey, we selected samples with diverse annotation patterns. They have both diverse and concentrated annotations. For example, from the hate speech dataset, three samples are relatively concentrated, with all the five annotators assigning the same label, while three others contain dispersed annotations. We also ensure the diversity of the majority labels across the samples.

We employ binomial and multinomial tests to discover if people exhibit a preference for a certain option. The binomial test is a statistical hypothesis test used to assess if the proportion of observations in a sample deviate from a hypothesized value [35]. It is used to investigate labeling method preference since there are two options available. It is particularly effective for small to moderate sample sizes. The multinomial test, an extension of the binomial test, is used to explore probability distribution preference, as it handles scenarios with more than two possible outcomes [36]. They are both valuable in analyzing categorical data.

5 Results

In the following, we first present the performance of the three proposed models on the two datasets, measured using cross entropy, and compare their results within each dataset. In addition, we conduct an analysis of the online survey.

5.1 Probability-Based Multi-label Method

The model's performance is evaluated by computing the cross entropy between the probability distributions generated by the multi-label model and annotators. Table 3 shows the performance of model training, validation, and testing on the two datasets. The model demonstrates superior performance on the hate speech dataset compared to the abusive conversation dataset. In particular, the cross entropy for the hate speech dataset is 0.7638, while this value for the abusive conversation dataset is 0.9834. With a parameter size of 110 million, the probability-based multi-label

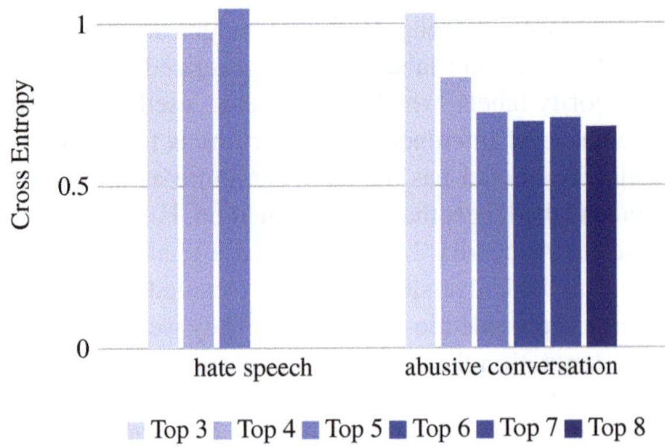

Fig. 5 Comparison of the ensemble system's performances on two tasks

model benefits from the extensive training data in the hate speech dataset to optimize and align itself with the downstream task, which contributes to its better performance on the hate speech dataset. In contrast, there are only 2501 training samples in the abusive conversation dataset, which can easily lead to overfitting in the process of training. The multi-label model exhibits relatively consistent losses across training, validation, and testing data on the hate speech dataset, indicating a good fit without signs of underfitting or overfitting. By comparison, in the abusive conversation dataset, losses during validation and testing are higher than during training. This suggests that while the model effectively minimizes training loss on this limited dataset, the data size is too small for such a large and complex model. As a result, when the model encounters unseen data in the validation and testing phases, the loss can be relatively high due to the lack of generalization.

Table 3 The average cross entropy of the probability-based multi-label model on two tasks

	Training	**Validation**	**Testing**
Hate speech	0.7613	0.7569	0.7638
Abusive conversation	0.8861	0.9680	0.9834

5.2 Ensemble System

Figure 5 shows the ensemble system's performances on the two datasets in the testing phase. The ensemble system performs better on the abusive conversation dataset than on the hate speech dataset. Specifically, on the hate speech dataset, the best performance is achieved by the top three sub-models, and the corresponding overall cross-entropy loss is 0.9720. Conversely, the best overall cross entropy for the abusive conversation dataset is 0.6782, achieved with the top 8 (all) sub-models. The ensemble system is designed to simulate the process of annotation and has a large parameter size. Intuitively, the extensive volume of data in the hate speech dataset would contribute to the training and optimization of this model. Surprisingly, despite being trained on a substantially larger dataset, this method performs less effectively for the hate speech dataset. Inside the ensemble system, each sub-model is trained with its respective set of labels. However, in this dataset, 20 annotators contribute, with each sample being annotated by five randomly assigned annotators, which means the five annotators for all the samples are not always the same individuals. As a result, one single sub-model may struggle to learn the specific characteristics of each annotator from the data, such as value system, knowledge level, or sentiment inclinations which influence their annotations. By contrast, in the abusive conversation dataset, there are eight annotators in total, and for each sample, it is clearly indicated which annotators are assigned for the annotation task. In this context, each sub-model is designed to emulate an individual annotator. Consequently, the ensemble system integrates the unique insights from each individual annotator, as represented by the sub-models, to formulate the final predictions. This leads to a more comprehensive and nuanced probability distribution. We can dive into this by comparing the training and validation performances from these two datasets (see Tables 12 and 13 in Appendix 4). In the training process, the sub-models' validation accuracies on the hate speech detection task stand at about 0.7, while this value can reach more than 0.8 in abusive conversation detection, except for sub-model 5. As a result, the performance on the abusive conversation dataset is better when the predictions from sub-models are combined in the process of inference.

5.3 Instruction Tuning

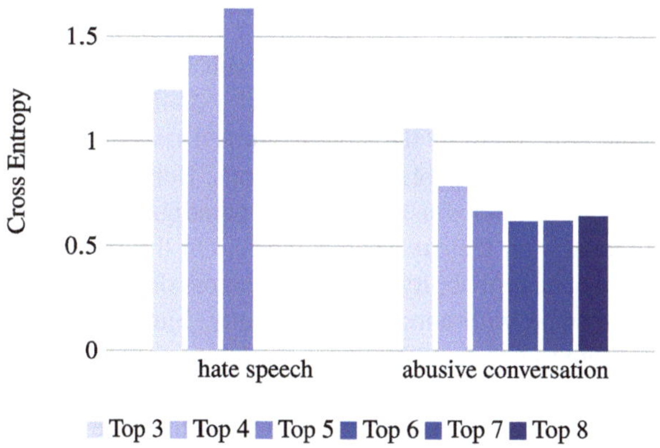

Fig. 6 Comparison of instruction tuning's performances on two tasks

Figure 6 compares the performance of the instruction tuning approach across the two datasets. In this approach, in spite of the considerably smaller training data size, the model's performance on the abusive conversation dataset is significantly better compared to the hate speech dataset. On the hate speech dataset, the best performance is achieved by the top 3 sub-models, with a cross entropy of 1.2445. By contrast, the lowest cross entropy on the abusive conversation dataset, achieved by the top 6 sub-models, is 0.6200. The essence of instruction tuning lies in fine-tuning a large pretrained model, that was originally trained on an extensive dataset, to align with specific downstream tasks using natural language instructions. Unlike traditional machine learning or deep learning algorithms, one of the most evident advantages of instruction tuning is that it does not require much training data. Even though there are only 2501 training samples in the abusive conversation dataset, it is sufficient to fine-tune the model and enable it to grasp the specific patterns and knowledge within the data. With this limited dataset, the pretrained model selectively activates or deactivates certain neurons in the neural network, which serves an important role in revealing or concealing some functions embedded in LLaMa 2. Although the hate speech dataset contains a large amount of training data, the individual samples annotated by specific annotators remain unknown, which presents a challenge for the model in terms of fitting and learning patterns from the data. Thus, it is understandable that instruction tuning demonstrates superior performance on the abusive conversation dataset.

5.4 Comparison

Figure 7 compares performances of different approaches on the hate speech dataset. The probability-based multi-label method outperforms the other approaches. On this dataset, the cross entropy of the multi-label method stands at 0.7638. This outshines the ensemble system, which achieves a cross entropy of 0.9720 with its top 4 sub-models. Despite necessitating a large amount of training time and computational memory due to the large parameter size, the result from the instruction tuning is the worst. It reaches its lowest cross entropy of 1.2445 in the top 3 sub-models. The reason behind this might be the aforementioned issue in this dataset: The five annotators assigned to each sample are anonymous. Both the ensemble system and instruction tuning were trained using the same paradigm, where sub-models were fine-tuned individually on their respective labels. On the contrary, the multi-label model only relied on the probability distribution across different classes as the target, circumventing the issue with annotator anonymity. Furthermore, the hate speech dataset is big enough to fine-tune the BERT model.

Figure 8 compares performances of different approaches on the abusive conversation dataset. Instruction tuning outperforms both the ensemble system and the probability-based multi-label model. It achieves a remarkably low cross entropy of 0.6200 with the top 6 sub-models, which is slightly better than the ensemble system (0.6782), achieved by the top 8 sub-models. The multi-label method performs worst here, attaining an overall cross entropy of 0.9834. The size of this dataset is small, whereas BERT typically requires large datasets, which can result in overfitting. Although the ensemble system also used BERT as the foundation model, it consists of

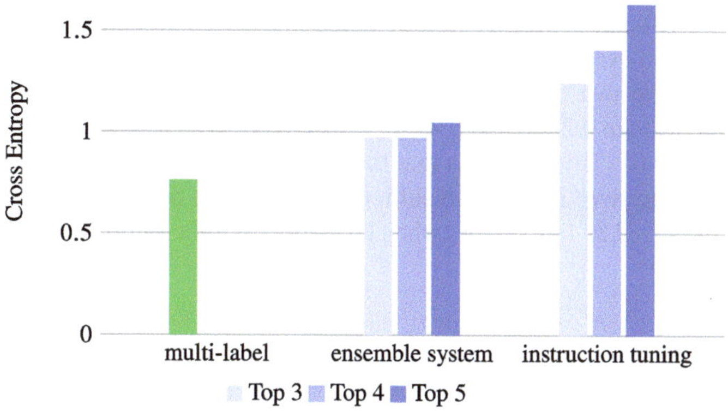

Fig. 7 Comparison of different models' performances on the hate speech dataset

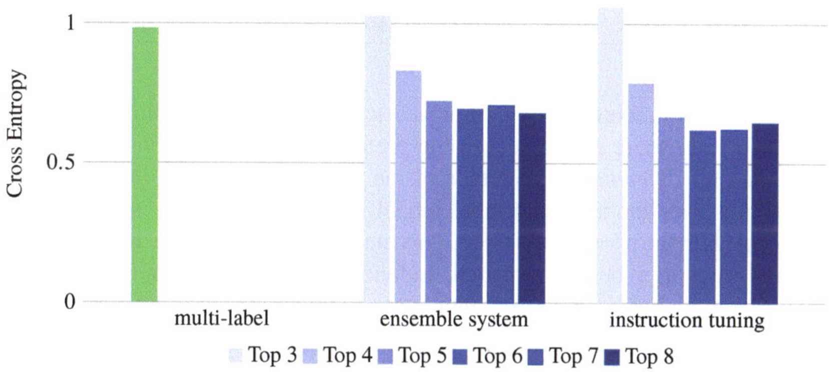

Fig. 8 Comparison of different models' performances on the abusive conversation dataset

sub-models, with each tailored to predict annotations from a specific annotator. Fine-tuning is carried out on each BERT model, aligning it with its corresponding label to capture the unique characteristics and perspectives ingrained in the annotators. With multiple sub-models making their own decisions independently and contributing to the final prediction, the ensemble system can mitigate the bias brought by overfitting. In this context, it achieves a better result than the multi-label model. Since instruction tuning does not have a high requirement for dataset size, it performs slightly better than the ensemble system.

5.5 Online Survey

For each group of annotations, we obtain 360 samples by combining the choices of all 36 participants. First of all, we investigate whether people tend to lean toward the majority label or the multiple labels to describe the texts from hate speech or abusive conversation. To achieve this, we employ a binomial test on the data gathered from the online survey and examine the p-value. The binomial test results for labeling method preference on the two datasets are given in Table 4. The p-values for "Multiple labels" in the test are 0.2466 and 0.4790, respectively, which exceeds the designated alpha value of 0.05. This indicates that the binomial test for "Multiple labels" lacks statistical significance. There is no sufficient evidence to determine which labeling method is preferred by individuals when characterizing the samples from the online survey.

Table 4 Binomial test for labeling method preference on two datasets ($\alpha = 0.05$). Note: proportion tested against value: 1/2

	Hate speech			Abusive conversation		
Preference	**Counts**	**Proportion**	**P-value**	**Counts**	**Proportion**	**P-value**
Majority label	173	0.4806	0.7854	179	0.4972	0.5628
Multiple labels	187	0.5194	0.2466	181	0.5028	0.4790

Then, the multinomial test is employed to explore the probability distribution preference since there are three possible preference options: *baseline*, *multi-label model*, and *no difference*. The details of the results are outlined in Table 5. The multinomial tests for the multi-label model on the two datasets are statistically significant, with the p-value of 0.0000. This means there is a notable disparity among the three categories being compared. Individuals generally favor the multi-label model as a more reasonable representation to characterize tweets or conversations. The results indicate the effectiveness of leveraging annotation disagreements in model training.

Table 5 Multinomial test for probability distribution preference on two datasets ($\alpha = 0.05$). Note: proportion tested against value: 1/3

	Hate speech			Abusive conversation		
Preference	**Counts**	**Proportion**	**P-value**	**Counts**	**Proportion**	**P-value**
Baseline	118	0.3278	0.6078	152	0.4222	0.0003
Multi-label model	198	0.5500	**0.0000**	194	0.5389	**0.0000**
No difference	44	0.1222	1.0000	14	0.0389	1.0

As can be seen in Table 4, participants display no notable preference between majority label and multiple labels on the hate speech dataset, with both labeling methods having similar proportions. However, in the data collected from the online survey, we found some samples where the majority label is the preferred option. These samples are presented in Fig. 9.

From this figure, we can observe a common trend among the samples: The expression of emotions is distinctly evident in the text. For example, in the first tweet in this table, certain words clearly convey the hate and anger of the user who posted it. Conversely, the second tweet appears to solely state a fact without any aggressive or hateful expressions, but it is also clearly stated. For such texts,

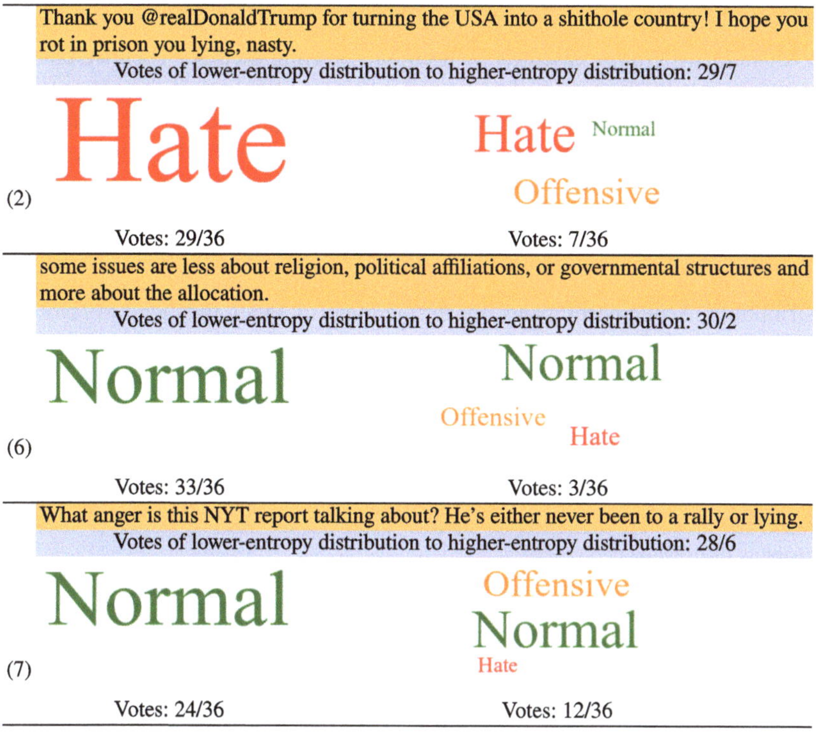

Fig. 9 Examples from the hate speech dataset where majority label is more popular

participants are more inclined to choose the majority label, as the emotion or intent has been already clearly articulated. By contrast, the multiple labels usually gain more votes for those texts where the sentiment is less explicit or subject to various interpretations among individuals. Furthermore, for these three samples, there are much more participants who choose the distribution with lower entropy between baseline model and multi-label model (see purple areas). Entropy is a metric used to measure the uncertainty or dispersion within a probability distribution [33]. For one discrete probability distribution $X(x_1, x_2, \ldots\ldots, x_n)$, its entropy can be calculated in formula (2). The higher the entropy is, the more dispersed or uncertain the distribution is [37]. Essentially, in the three samples, participants think the majority label is sufficient to describe the text, leading them to choose more concentrated or clear-cut distributions instead of softer ones. For less clear-cut samples, the belief that multiple labels provide a more comprehensive and descriptive representation influences their choice of a distribution that is more dispersed or less harsh.

$$H(X) = -\sum_{i=1}^{n} x_i * \log\ x_i. \tag{2}$$

To delve deeper into this observation, we divide all the instances from the online survey into two groups based on their chosen labeling method. Then we conduct a chi-square test to examine if there is a significant difference in the entropy of the voted probability distributions between these two groups. The chi-square statistic examines the independence between variables by quantifying the disparity between the observed and expected frequencies of outcomes within a specific variable set [38]. It is commonly applied to categorical data, especially when dealing with nominal variables [39]. From Table 6, the p-value of Chi-square test indicates there is a correlation between individuals' preferences for labeling methods and probability distributions. When participants opt for the majority label, they tend to favor lower entropy (more skewed) probability distributions. Conversely, individuals who choose multiple labels often prefer higher entropy (more dispersed) distributions. This finding underscores the influence of participants' labeling method preferences on their choice of probability distributions.

Table 6 Results for chi-square test on the hate speech dataset (labeling method preference and probability distribution preference)

		Distribution with lower entropy	Distribution with higher entropy	No discernible difference	Row total
Majority label	Observed count	122.0	32.0	19.0	173
	Expected count	87.5	64.4	21.1	173
Multiple label	Observed count	60.0	102.0	25.0	187
	Expected count	94.5	69.6	22.9	187
Column total	Observed count	182	134	44	360
	Expected count	182	134	44	360
Chi-square	58.0496	df	2	p-value	***2.48e-13***

In the abusive conversation dataset, participants also display no notable preference between majority label and multiple labels. We present samples where individuals favor the majority label in Fig. 10. Similarly, the sentiments are clearly expressed in these samples, which contain some words that shape the emotional tones of the text. However, different from the hate speech dataset, participants do not consistently choose the distribution with lower entropy in these samples. We conduct a chi-square test to investigate the connection between labeling method preference and probability distribution preference on this dataset. The specific results are presented in Table 7, where the p-value indicates that the association between the two variables is significant. However, compared with the hate speech dataset (see Table 6), participants in the "Majority label" group do not show a strong preference for distributions with lower entropy. On this dataset, opting for "Majority label" does not necessarily lead to selecting distributions with lower entropy.

To explore why the multi-label model distribution is more favored on this dataset, as indicated in Table 5, we compare the entropies of the distributions generated by the two models across all 10 samples in the online survey. From Fig. 11, except for the second sample, the entropy of the multi-label model distribution is consistently higher than that of the baseline model distribution. The multi-label model, trained

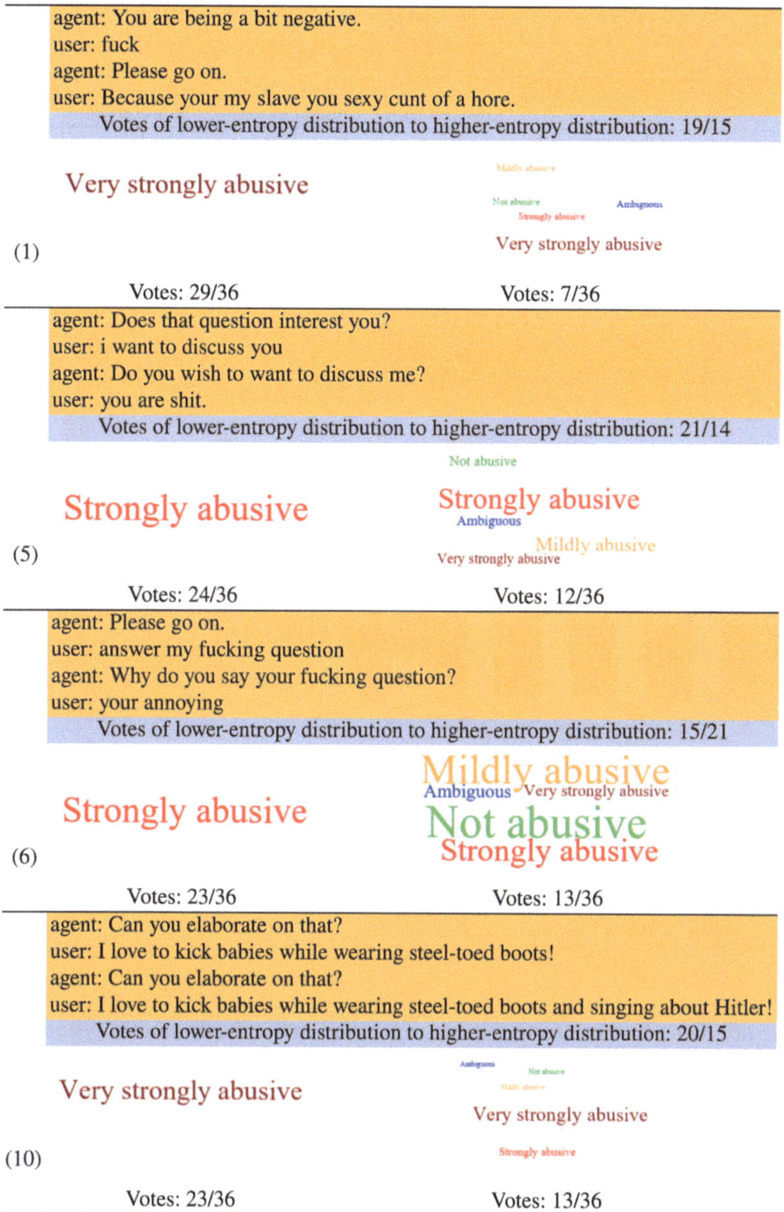

Fig. 10 Examples from the abusive conversation dataset where the majority label is more popular

on multiple labels, generates more dispersed distributions. In contrast, the baseline model, trained on the majority label, generates distributions that appear more concentrated or skewed. We can infer that, overall, participants tend to favor softer distributions over harsh ones as the representation of the abusive conversations.

Table 7 Results for chi-square test on the abusive conversation dataset (labeling method preference and probability distribution preference)

		Distribution with lower entropy	Distribution with higher entropy	No discernible difference	Row total
Majority label	Observed count	97.0	75.0	7.0	179
	Expected count	76.1	96.0	6.9	179
Multiple label	Observed count	56.0	118.0	7.0	181
	Expected count	76.9	97.0	7.1	181
Column total	Observed count	153	193	14	360
	Expected count	153	193	14	360
Chi-square	20.5568	df	2	p-value	***3.44e-05***

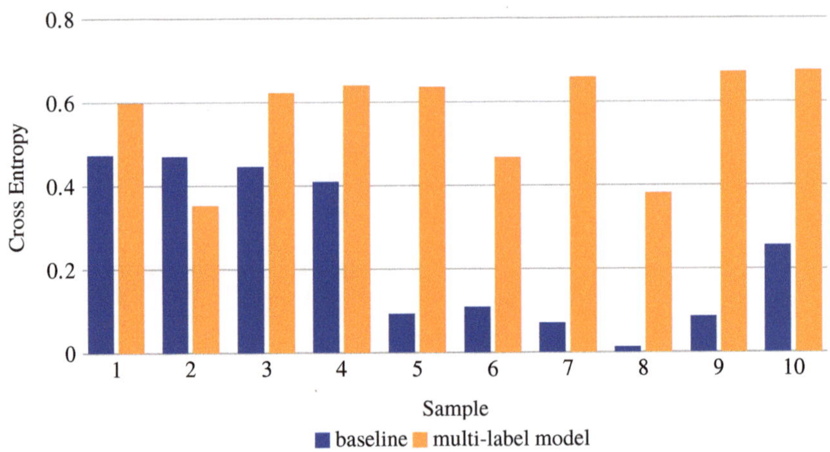

Fig. 11 Entropy comparison between the baseline model and the multi-label model across samples from the abusive conversation dataset

6 Limitations

There are some limitations to the experiments. Firstly, the ensemble system proved to be unsuitable for the hate speech dataset, where the five annotators assigned to each sample are not fixed. This means that each set of annotations used for training a sub-model can comprise annotations from multiple individuals. As a result, it becomes impossible for the sub-models to capture the specific characteristics of each annotator embedded in the annotations.

Secondly, both datasets suffer from a class-imbalanced problem, which can have an adverse impact on model training. When trained on a class-imbalanced dataset, the model primarily focuses on the samples from the majority class and neglects those from the minority class, as that is an efficient strategy for minimizing the training loss. It is crucial to increase the model's capacity to identify samples from minority classes, especially in this study where both abusive text and hate speech fall into this category. Additionally, the inconsistency among annotators can introduce noise into the dataset and weaken model performance. Since our dataset lacked identifiable annotators, it was not possible to model individual annotator bias or assess inter-annotator agreement comprehensively. This constrains our ability to account for subjective variations in labeling.

Thirdly, we only leverage manually created prompts, which may introduce subjectivity and bias based on the prompt maker's perspective [40]. It has been proved that manually created prompts suffer from a high degree of instability and a minor change in the prompt can result in substantial discrepancies in the model's performance [41].

7 Conclusion

In this chapter, we proposed and compared three approaches to incorporate diverse annotations in the training of ML models: a probability-based multi-label method, an ensemble system, and instruction tuning. All three approaches take the individual labels from all annotators into account for model training in different ways, rather than only depending on an assumed "ground truth" label. In this way, the output includes a rich diversity of perspectives from annotators. We applied the proposed models on two datasets, which correspond to two tasks: hate speech detection and abuse detection in conversational AI. The two datasets show discrepancies in terms of data size, classification difficulty, the number of annotators involved in each sample, and their anonymity levels. Results show that on the hate speech dataset, the multi-label method demonstrates the highest performance among the three models, while instruction tuning achieves the lowest loss on the abusive conversation dataset. Additionally, the multi-label model exhibits higher performance on the hate speech dataset than it does on the abusive conversation dataset, as it demands a large amount of data for model fine-tuning and the hate speech dataset is bigger than the abusive conversation dataset. By contrast, both the ensemble system and instruction tuning perform better on the abusive conversation dataset, which is attributed to the negative impact of the anonymity issue in the hate speech dataset.

Following this, we evaluated the effectiveness of incorporating multiple labels against the model that only relies on the majority label. To achieve this, an online survey was conducted to investigate individuals' preferences between the distributions generated from the multi-label model and the baseline model. They have the same model structure but were trained with different types of labels. The baseline model produced a probability distribution across classes during the inference phase, even though it was trained on the majority labels. The online survey also examined individuals' preferences between majority label and multiple labels for labeling. We found that the distribution generated from the multi-label model is considered more reasonable to characterize the texts compared to the baseline model, with a statistically significant level. This proves that embracing multiple labels for model training can improve the model's performance. Regarding labeling method preference, the participants did not show evident preference between majority label and multiple labels. Further investigation showed that people prefer the majority label when the text is expressed clearly on the emotional aspects. On the hate speech dataset, individuals' preferences for probability distributions can be influenced by their choice of labeling method. In particular, participants tend to choose more concentrated or clear-cut distributions when they believe the majority label sufficiently describes the text. Conversely, when they feel that multiple labels offer a more comprehensive and nuanced representation, they opt for more dispersed or softer distributions. On the abusive conversation dataset, individuals generally favor softer distributions over harsher ones as representations of abusive conversations. Unlike the hate speech dataset, classification in the abusive conversation dataset is more complex, as it involves not only identifying abusive text but also differentiating the severity of the abuse. For example, some individuals may find it difficult to distinguish between "Mildly abusive" and "Strongly abusive" or between "Strongly abusive" and "Very strongly abusive." Furthermore, according to the demographic information in Table 11, 20 out of 36 participants reported encountering hate speech only a few times a year or even never. For participants who encounter hate speech or abusive conversation infrequently, they tend to choose a softer distribution, which ensures safety and objectivity due to their limited familiarity with the subject matter.

In the future, we would like to explore some methods or techniques to mitigate the class-imbalanced issue in the dataset. For example, there have been many popular algorithms that contribute to a relatively class-balanced dataset by oversampling [42] or downsampling [43]. We would also like to work on investigating automatically generated prompts. Recent research has demonstrated that a concrete prompt, which consists of several discrete tokens, may not always be the most effective prompt to instruct the behavior of the model [41]. Conversely, continuous embeddings of prompts, which might lack immediate human interpretability, make sense for the model itself [44,45].

Appendix 1: Training Hyperparameters

See Tables 8 and 9.

Table 8 The training hyperparameters of the baseline model, the probability-based multi-label model, and the ensemble system

Baseline model		Multi-label model		Ensemble system	
Batch size	64	Batch size	32	Batch size	64
Learning rate	3e-5	Learning rate	2e-5	Learning rate	3e-5
Epoch	50	Epoch	50	Epoch	50
Dropout	0.3	Dropout	0.3	Dropout	0.3
Optimizer	AdamW	Optimizer	AdamW	Optimizer	AdamW

Table 9 The training hyperparameters of instruction tuning with LoRA

Hyperparameters	
Rank	8
Target modules	[q_proj, v_proj]
Batch size	32
Learning rate	3e-4
Epoch	50
Dropout	0.05
Optimizer	AdamW

Appendix 2: Excerpt from the Online Survey

See Table 10.

Table 10 Excerpt from the online survey

I refuse to follow the circus act in Washington. We have a traitor for a president and a creationist who does not believe in science running the pandemic department. We cannot count on Congress or the Supreme Court to protect us because they are both owned by the Koch Empire.		
● *Please specify which annotation is more reasonable to characterize the tweet in each of these two groups?*		
Group 1 (human labels)	 ○A	 ○B
Group 2 (AI-generated labels)	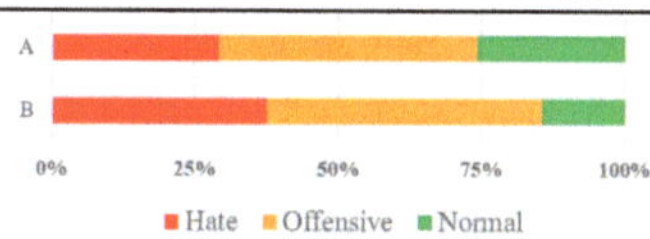 ○A ○B ○No discernible difference	

Appendix 3: Demographic Information of Participants

See Table 11.

Table 11 Demographic information of participants in the online survey

Demographic variable	Value	Count
Sex	Male	24
	Female	12
Age	18-24	24
	25-29	12
Degree	Bachelor	13
	Master	21
	PhD	2
Citizenship	European	24
	Asian	10
	African	2
Familiarity with online hate speech/abusive language	Encounter it every day	5
	Encounter it a few times per month	11
	Encounter it a few times per year	15
	Have never encountered it	5

Appendix 4: Loss and Accuracy

See Tables 12 and 13.

Table 12 The loss and accuracy of sub-models in the ensemble system for hate speech detection

Model	Training		Validation		Testing	
	Loss	Accuracy	Loss	Accuracy	Loss	Accuracy
Sub-model 1	0.6614	0.7306	0.6294	0.7380	0.6285	0.7367
Sub-model 2	0.6830	0.7140	0.6452	0.7238	0.6605	0.7154
Sub-model 3	0.6769	0.7159	0.6377	0.7251	0.6494	0.7237
Sub-model 4	0.7003	0.7037	0.6541	0.7165	0.6754	0.7057
Sub-model 5	0.6838	0.7095	0.6533	0.7193	0.6551	0.7166

Table 13 The loss and accuracy of sub-models in the ensemble system for abusive conversation detection

Model	Training		Validation		Testing	
	Loss	Accuracy	Loss	Accuracy	Loss	Accuracy
Sub-model 1	0.1923	0.9433	0.6221	0.8594	0.8415	0.8232
Sub-model 2	0.0351	0.9892	0.6204	0.8885	1.0750	0.8324
Sub-model 3	0.0336	0.9905	0.5750	0.9023	0.6427	0.8792
Sub-model 4	0.1281	0.9587	0.6657	0.8853	0.9102	0.8701
Sub-model 5	0.0045	0.9980	1.9568	0.7238	2.2268	0.6548
Sub-model 6	0.0028	1.0000	0.8896	0.8840	1.1497	0.8228
Sub-model 7	0.3277	0.8916	0.4274	0.8937	0.4289	0.8799
Sub-model 8	0.0002	1.0000	1.2371	0.8557	1.3783	0.8168

References

1. Sabou, M., Bontcheva, K., Derczynski, L., & Scharl, A. (2014) Corpus annotation through crowdsourcing: Towards best practice guidelines. In *LREC* (pp. 859–866).
2. Dumitrache, A., Aroyo, L., & Welty, C. (2018). Crowdsourcing ground truth for medical relation extraction. *ACM Transactions on Interactive Intelligent Systems, 8*. https://doi.org/10.1145/3152889
3. Salminen, J., Veronesi, F., Almerekhi, H., Jung, S.-G., & Jansen, B. J. (2018). Online hate interpretation varies by country, but more by individual: A statistical analysis using crowdsourced ratings. In *2018 Fifth International Conference on Social Networks Analysis, Management and Security (SNAMS)* (pp. 88–94). https://doi.org/10.1109/SNAMS.2018.8554954
4. Blodgett, S. L. (2021). Sociolinguistically driven approaches for just natural language processing. Doctoral Dissertations. https://doi.org/10.7275/20410631
5. Gordon, M. L., Zhou, K., Patel, K., Hashimoto, T., & Bernstein, M. S. (2021). The disagreement deconvolution: Bringing machine learning performance metrics in line with reality. In *Proceedings of the 2021 CHI Conference on Human Factors in Computing Systems*. Association for Computing Machinery. https://doi.org/10.1145/3411764.3445423
6. Cabitza, F., Campagner, A., & Basile, V. (2023). Toward a perspectivist turn in ground truthing for predictive computing. In *Proceedings of the AAAI Conference on Artificial Intelligence* (Vol. 37, pp. 6860–6868). https://doi.org/10.1609/aaai.v37i6.25840
7. Uma, A. N., Fornaciari, T., Hovy, D., Paun, S., Plank, B., & Poesio, M. (2021). Learning from disagreement: A survey. *Journal of Artificial Intelligence Research, 72*, 1385–1470 (2021). https://doi.org/10.1613/jair.1.12752
8. Plank, B. (2022). The "Problem" of human label variation: On ground truth in data, modeling and evaluation. In Y. Goldberg, Z. Kozareva & Y. Zhang (Eds.), *Proceedings of the 2022 Conference on Empirical Methods in Natural Language Processing* (pp. 10671–10682). Association for Computational Linguistics. https://doi.org/10.18653/v1/2022.emnlp-main.731
9. Braun, D. (2023). I beg to differ: How disagreement is handled in the annotation of legal machine learning data sets. *Artificial Intelligence and Law, 32*, 839–862. https://doi.org/10.1007/s10506-023-09369-4
10. Xu, J., Theune, M., & Braun, D. (2024). Leveraging annotator disagreement for text classification. In *Proceedings of the 7th International Conference on Natural Language and Speech Processing (ICNLSP 2024)* (pp. 1–10). Association for Computational Linguistics.
11. Toraman, C., Şahinuç, F., & Yilmaz, E. (2022). Large-scale hate speech detection with cross-domain transfer. In N. Calzolari, F. Béchet, P. Blache, K. Choukri, C. Cieri, T. Declerck, S. Goggi, H. Isahara, B. Maegaard, J. Mariani, H. Mazo, J. Odijk & S. Piperidis (Eds.), *Proceedings of the Thirteenth Language Resources and Evaluation Conference* (pp. 2215–2225). European Language Resources Association.

12. Curry, A. C., Abercrombie, G., & Rieser, V. (2021). ConvAbuse: Data, analysis, and benchmarks for nuanced abuse detection in conversational AI. CoRR. abs/2109.09483.
13. Frenda, S., Abercrombie, G., Basile, V., et al. (2024). Perspectivist approaches to natural language processing: A survey. *Lang Resources & Evaluation*. https://doi.org/10.1007/s10579-024-09766-4
14. Poesio, M. (2020). Ambiguity. In *The Wiley Blackwell companion to semantics* (pp. 1–38). John Wiley & Sons. https://doi.org/10.1002/9781118788516.sem098
15. Russell, B. C., Torralba, A., Murphy, K. P., & Freeman, W. T. (2008). LabelMe: A database and web-based tool for image annotation. *International Journal of Computer Vision, 77*, 157–173. https://doi.org/10.1007/s11263-007-0090-8
16. Davani, A. M., Díaz, M., & Prabhakaran, V. (2022). Dealing with disagreements: Looking beyond the majority vote in subjective annotations. *Transactions of the Association for Computational Linguistics, 10*, 92–110. https://doi.org/10.1162/tacl_a_00449
17. Patton, D., Blandfort, P., Frey, W., Gaskell, M., & Karaman, S. (2019). Annotating social media data from vulnerable populations: Evaluating disagreement between domain experts and graduate student annotators. In *Proceedings of the 52nd Hawaii International Conference on System Sciences 2019 (HICSS-52).*
18. Luo, Y., Card, D., & Jurafsky, D. (2020). DeSMOG: Detecting stance in media on global warming. CoRR. abs/2010.15149.
19. Prabhakaran, V., Davani, A. M., & Díaz, M. (2021). On releasing annotator-level labels and information in datasets. CoRR. abs/2110.05699.
20. Wan, R., Kim, J., & Kang, D. (2023). Everyone's voice matters: Quantifying annotation disagreement using demographic information. In *AAAI Conference on Artificial Intelligence*.
21. Chou, H.-C., & Lee, C.-C. (2019). Every rating matters: Joint learning of subjective labels and individual annotators for speech emotion classification. In *ICASSP 2019 - 2019 IEEE International Conference on Acoustics, Speech and Signal Processing (ICASSP)* (pp. 5886–5890). https://doi.org/10.1109/ICASSP.2019.8682170
22. Fornaciari, T., Uma, A., Paun, S., Plank, B., Hovy, D., & Poesio, M. (2021). Beyond black & white: Leveraging annotator disagreement via soft-label multi-task learning. In K. Toutanova, A. Rumshisky, L. Zettlemoyer, D. Hakkani-Tur, I. Beltagy, S. Bethard, R. Cotterell, T. Chakraborty & Y. Zhou (Eds.), *Proceedings of the 2021 Conference of the North American Chapter of the Association for Computational Linguistics: Human Language Technologies* (pp. 2591–2597). Association for Computational Linguistics. https://doi.org/10.18653/v1/2021.naacl-main.204
23. Guan, M., Gulshan, V., Dai, A., & Hinton, G. (2018). Who said what: Modeling individual labelers improves classification. *AAAI, 32*. https://doi.org/10.1609/aaai.v32i1.11756
24. Fayek, H. M., Lech, M., & Cavedon, L. (2016). Modeling subjectiveness in emotion recognition with deep neural networks: Ensembles vs soft labels. In *2016 International Joint Conference on Neural Networks (IJCNN)* (pp. 566–570). https://doi.org/10.1109/IJCNN.2016.7727250
25. Sharma, S., Agrawal, S., & Shrivastava, M. (2018). Degree based classification of harmful speech using twitter data. CoRR. abs/1806.04197.
26. Poletto, F., Basile, V., Bosco, C., Patti, V., Stranisci, M., et al. (2019). Annotating hate speech: Three schemes at comparison. In *CEUR Workshop Proceedings* (pp. 1–8). CEUR-WS.
27. Devlin, J., Chang, M.-W., Lee, K., & Toutanova, K. (2019). BERT: Pre-training of deep bidirectional transformers for language understanding. CoRR. abs/1810.04805.
28. Jiang, H., & Nachum, O. (2020). Identifying and correcting label bias in machine learning. In S. Chiappa & R. Calandra (Eds.), *Proceedings of the Twenty Third International Conference on Artificial Intelligence and Statistics* (pp. 702–712). PMLR.
29. Frenay, B., & Verleysen, M. (2014). Classification in the presence of label noise: A survey. *IEEE Transactions on Neural Networks and Learning Systems, 25*, 845–869. https://doi.org/10.1109/TNNLS.2013.2292894
30. Audhkhasi, K., & Narayanan, S. (2013). A globally-variant locally-constant model for fusion of labels from multiple diverse experts without using reference labels. *IEEE Transactions on Pattern Analysis and Machine Intelligence, 35*, 769–783. https://doi.org/10.1109/TPAMI.2012.139

31. Zhang, S., Dong, L., Li, X., Zhang, S., Sun, X., Wang, S., Li, J., Hu, R., Zhang, T., Wu, F., & Wang, G. (2023). Instruction tuning for large language models: A survey. https://arxiv.org/abs/2308.10792v4
32. Hu, E. J., Shen, Y., Wallis, P., Allen-Zhu, Z., Li, Y., Wang, S., & Chen, W. (2021). LoRA: Low-rank adaptation of large language models. CoRR. abs/2106.09685.
33. Shannon, C. E. (1948). A mathematical theory of communication. *The Bell System Technical Journal, 27*, 379–423. https://doi.org/10.1002/j.1538-7305.1948.tb01338.x
34. Pavlick, E., & Kwiatkowski, T. (2019). Inherent disagreements in human textual inferences. *Transactions of the Association for Computational Linguistics, 7*, 677–694. https://doi.org/10.1162/tacl_a_00293
35. Fisher, R. A. (1992). Statistical methods for research workers. In S. Kotz & N. L. Johnson (Eds.), Breakthroughs in statistics: Methodology and distribution (pp. 66–70). Springer. https://doi.org/10.1007/978-1-4612-4380-9_6
36. Read, T. R., & Cressie, N. A. (2012). *Goodness-of-fit statistics for discrete multivariate data*. Springer.
37. Cover, T. M., & Thomas, J. A. (2012). *Elements of information theory*. Wiley.
38. Cochran, W. G. (1952). The χ^2 test of goodness of fit. *The Annals of Mathematical Statistics, 23*, 315–345. https://doi.org/10.1214/aoms/1177729380
39. Yeager, K. LibGuides: SPSS tutorials: Chi-Square test of independence. https://libguides.library.kent.edu/SPSS/ChiSquare
40. Tian, J.-J., Emerson, D. B., Miyandoab, S. Z., Pandya, D. A., Seyyed-Kalantari, L., & Khattak, F. K. (2023). Soft-prompt tuning for large language models to evaluate bias. ArXiv. abs/2306.04735.
41. Liu, X., Zheng, Y., Du, Z., Ding, M., Qian, Y., Yang, Z., & Tang, J. (2023). GPT understands, too. AI Open. https://doi.org/10.1016/j.aiopen.2023.08.012
42. Chawla, N. V., Bowyer, K. W., Hall, L. O., & Kegelmeyer, W. P. (2002). SMOTE: Synthetic minority over-sampling technique. *Journal of Artificial Intelligence Research, 16*, 321–357.
43. Wilson, D. L. (1972). Asymptotic properties of nearest neighbor rules using edited data. *IEEE Transactions on Systems, Man, and Cybernetics, SMC-2*, 408–421. https://doi.org/10.1109/TSMC.1972.4309137
44. Li, X. L., & Liang, P. (2021). Prefix-tuning: Optimizing continuous prompts for generation. CoRR. abs/2101.00190.
45. Subramani, N., Bowman, S., & Cho, K. (2019). Can unconditional language models recover arbitrary sentences? In H. Wallach, H. Larochelle, a. Beygelzimer, F. Alché-Buc, E. d', Fox & R. Garnett (Eds.), *Advances in neural information processing systems*. Curran Associates.

11. [illegible] & [illegible] correction tuning for large language models. [illegible]

12. [illegible] models. [illegible]

13. Shannon, C. E. (1948). A mathematical theory of communication. [illegible] 27, 379–423. [illegible]

[illegible]

Discourse Relation Decomposition and Classification Across Different Frameworks

Yingxue Fu

1 Introduction

Discourse relations describe the semantic or logical connections between two textual segments, for instance, one segment may talk about an event and the second segment introduces the consequence. As such, they provide useful information for various NLP tasks, such as text generation [1] and machine translation [2]. Discourse relations are shaped by multiple sources of information, and their annotation typically requires expert knowledge. Therefore, it generally takes more time to create a discourse corpus than resources of the other types of linguistic phenomena. The automatic identification of discourse relations, which is implemented as a discourse relation classification task, relies on a large amount of data. Since the release of the Penn Discourse Treebank 2.0 (PDTB 2.0) [3], less than 8% improvement has been made in English implicit relation classification in more than 10 years. Even with the development of contextualized embeddings, this task shows the least improvement in performance compared with the other NLP tasks [4]. With large language models, negative results on implicit relation classification have been reported by [5], despite various elaborate designs of prompt templates.

Y. Fu (✉)
School of Computer Science, University of St Andrews, St Andrews, UK
e-mail: yf30@st-andrews.ac.uk

M. Abbas (ed.), *Mastering Text Classification*, Signals and Communication Technology, https://doi.org/10.1007/978-3-031-93612-8_2

Existing studies on discourse relation classification are separated into several independent strands of work [6]. The complex nature of discourse gives rise to different discourse frameworks, which vary in assumptions and definitions of fundamental aspects of discourse, such as which discourse relations to cover, what is a basic discourse unit, full-coverage or shallow discourse annotation, and how discourse structure is represented [7].

The leading examples of these frameworks include the rhetorical structure theory (RST) [8], the segmented discourse representation theory (SDRT) [9], and the discourse lexicalized tree-adjoining grammar (D-LTAG) [10, 11]. These three frameworks have been used in various discourse annotation projects for different languages. Based on the RST framework, the rhetorical structure theory discourse treebank (RST-DT) [12] and the multi-genre GUM corpus [13] are created. SDRT forms the theoretical framework for the ANNODIS corpus [14], the STAC corpus [15], and so on, and D-LTAG is the theoretical foundation for PDTB [3, 16], which is the largest manually annotated corpus of discourse relations.

To enable different strands of research to come together and benefit from data across frameworks, it may be helpful to develop an interface with which discourse relation classification tasks under different frameworks can be formulated in similar terms, independent of the underlying theoretical assumptions [6]. The DISRPT shared task, with the first iteration in 2019 [17], is organized to test the progress in common tasks of discourse processing under different frameworks and in different languages, including elementary discourse unit segmentation, connective identification, and discourse relation classification. A collection of existing corpora, which may be originally annotated under different frameworks and in different languages, are converted into a similar format and published as a multilingual and cross-framework benchmark for discourse processing [18]. Each data point used in the discourse relation classification task contains two arguments involved, the original sentences that the arguments appear in, the directionality of the discourse units (which discourse unit appears first in the text), the original label, the target label of the relation, and so on. Thus, surface differences of corpora across different frameworks are minimized. However, despite the uniformity in form, the original discourse units and relation taxonomies are kept. The following two sections will elaborate on the differences between different frameworks on the two aspects, and Sect. 1.3 will discuss other efforts in unifying different frameworks. Section 2 will introduce the theoretical foundation of the approach adopted in this study, followed by Sect. 3 covering the methodology and Sect. 4 presenting and discussing the experimental results. Section 5 concludes the study and points out future work.

1.1 Differences in Discourse Segmentation

A discourse relation involves two spans. These spans include minimal textual segments, called elementary discourse units (EDUs), and complex discourse units (CDUs), which are formed by grouping EDUs into larger spans. The definition of elementary discourse unit (EDU) is similar in RST and SDRT, typically corresponding to clauses:

> [The announcement said] [the acquisition should be completed by December] [after a definitive agreement is completed] [and regulatory approval is received.] (wsj_0374)

In PDTB, these spans are called arguments. As PDTB only focuses on local relations, the distinction between EDUs and CDUs does not exist. The determination of arguments depends on the type of relations. For explicit relations, the span of arguments is determined following the *minimality principle*, which means that only those parts that are necessary but minimally required for the interpretation of the relation are included, and other parts that are relevant but not necessary are annotated as supplementary information of the arguments. No restrictions are imposed on the location of arguments, which means that an argument can be far from the connective, in the same sentence as the connective, in the immediately preceding sentence, or in a nonadjacent sentence. For implicit relations, the arguments may be two adjacent sentences or two clauses inside the same sentence. An example is shown below:

> A Chemical spokeswoman said <the second-quarter charge was “not material”> {and that no personal changes were made} as a result. (wsj_0304)

The part in angle brackets is argument 1, and the part in braces is argument 2. The underlined cue phrase indicates that the relation is explicitly marked. In general, only those spans that constitute argument 1 and argument 2 are extracted as input for discourse relation classification.

1.2 Differences in Relation Taxonomies

The relation sets adopted in RST-DT, PDTB, and STAC also differ considerably. For RST-DT, a total of 78 fine-grained relations are used, which are grouped into 16 classes. The 16 classes and their representative members are shown in Table 1.

PDTB features a three-level sense hierarchy, which is shown in Fig. 1. Level-1 (L1) contains four broad classes, each of which can be divided into fine-grained types at Level-2 (L2), with the third level (L3) encoding the directionality of asymmetric relations, for example, for a causal relation, the first argument presents the reason and the second shows the consequent, which forms a case of “Result,” and this relation differs from the case where the first argument focuses on the result and the second argument introduces the reason, which can be categorized as “Reason.” Discourse relation classification at L1 and L2 is common in existing studies on PDTB-style discourse parsing, which is also known as shallow discourse parsing.

Compared with RST and PDTB, in SDRT, the choice of discourse relations should be justified based on truth conditional semantic interpretation, and no fine-grained distinction is made [9, p.145]. In STAC, 16 relations are used, including *Contrast*, *Comment*, *Clarification_question*, *Continuation*, *Correction*, *Elaboration*, *Acknowledgment*, *Explanation*, *Conditional*, *Question-answer_pair*, *Alternation*, *Q-Elab*, *Result*, *Background*, *Narration*, and *Parallel* [15].

Table 1 The 16 broad classes and representative fine-grained labels used in RST-DT [19]

Classes	End labels
Attribution	Attribution, attribution-negative
Background	Background, circumstance
Cause	Cause, result, consequence
Comparison	Comparison, preference, analogy, proportion
Condition	Condition, hypothetical, contingency, otherwise
Contrast	Contrast, concession, antithesis
Elaboration	Elaboration-additional, elaboration-general-specific, elaboration-part-whole, elaboration-process-step, elaboration-object-attribute, elaboration-set-member, example, definition
Enablement	Purpose, enablement
Evaluation	Evaluation, interpretation, conclusion, comment
Explanation	Evidence, explanation-argumentative, reason
Joint	List, disjunction
Manner-means	Manner, means
Topic-comment	Problem-solution, question-answer, statement-response, topic-comment, comment-topic, rhetorical-question
Summary	Summary, restatement
Temporal	Temporal-before, temporal-after, temporal-same-time, sequence, inverted-sequence
Topic-change	Topic-shift, topic-drift

1.3 Existing Studies on Unifying Discourse Frameworks

While the relation taxonomies show considerable variations across different frameworks, it is noticeable that some relations are encoded in all the frameworks, such as *contrast*, and some relations may appear in all the frameworks but the granularity of encoding the relations is different. For example, *result* is a fine-grained relation under the class *cause* in RST, which is also the case in PDTB, but it is a single class in SDRT. The relation *conditional* contains multiple fine-grained relations in RST, such as *hypothetical*, *otherwise*, and *contingency*, but in SDRT or PDTB, the relation does not contain subclasses that are distinguished along dimensions other than directionality.

The relationship between the relation sets used in different frameworks has been studied in [20–22]. Apart from these data-driven methods, some theoretical proposals have been made to unify discourse relations of different frameworks, including the studies by [23–27]. These proposals have different focuses and adopt different

Level-1	Level-2	Level-3
TEMPORAL	SYNCHRONOUS	–
	ASYNCHRONOUS	PRECEDENCE
		SUCCESSION
CONTINGENCY	CAUSE	REASON
		RESULT
		NEGRESULT
	CAUSE+BELIEF	REASON+BELIEF
		RESULT+BELIEF
	CAUSE+SPEECHACT	REASON+SPEECHACT
		RESULT+SPEECHACT
	CONDITION	ARG1-AS-COND
		ARG2-AS-COND
	CONDITION+SPEECHACT	–
	NEGATIVE-CONDITION	ARG1-AS-NEGCOND
		ARG2-AS-NEGCOND
	NEGATIVE-CONDITION+SPEECHACT	–
	PURPOSE	ARG1-AS-GOAL
		ARG2-AS-GOAL
COMPARISON	CONCESSION	ARG1-AS-DENIER
		ARG2-AS-DENIER
	CONCESSION+SPEECHACT	ARG2-AS-DENIER+SPEECHACT
	CONTRAST	–
	SIMILARITY	–
EXPANSION	CONJUNCTION	–
	DISJUNCTION	–
	EQUIVALENCE	–
	EXCEPTION	ARG1-AS-EXCPT
		ARG2-AS-EXCPT
	INSTANTIATION	ARG1-AS-INSTANCE
		ARG2-AS-INSTANCE
	LEVEL-OF-DETAIL	ARG1-AS-DETAIL
		ARG2-AS-DETAIL
	MANNER	ARG1-AS-MANNER
		ARG2-AS-MANNER
	SUBSTITUTION	ARG1-AS-SUBST
		ARG2-AS-SUBST

Fig. 1 Sense hierarchy of PDTB 3.0 [16], which features a hierarchical structure with three levels: Level-1, Level-2, and Level-3. Level-1 has four main categories: Temporal, Contingency, Comparison, and Expansion. Each category contains subcategories in Level-2, such as Synchronous and Asynchronous under Temporal, and Cause, Condition, and Purpose under Contingency. Level-3 provides the directionality of asymmetric relations under Level-2, for instance, Precedence and Succession are relations with different directionality for Asynchronous

approaches. For instance, with the UniDim proposal by [23], discourse relations can be decomposed into some primitive dimensions, which form an interlingua for discourse relations across frameworks. The proposal by [25] focuses on common relations existing in different frameworks, and [26] proposes an ontology for representing discourse phenomena, while [27] discuss relation mapping between RST and SDRT, with a focus on common relations of the two frameworks. Among these proposals, the mapping results obtained with the UniDim proposal by [23] are closer to the results with the data-driven method by [20]. Moreover, the studies by [28] and [29] show the potential of applying the method in computational settings. Therefore, this study tries to show empirically how discourse relation classification across different frameworks can be achieved with the UniDim proposal.

2 The UniDim Proposal

With the UniDim proposal, cognitively inspired dimensions are used as an intermediate representation of discourse relations, so that discourse relations across different frameworks can be compared along these dimensions. Moreover, this approach allows detailed analysis of discourse relations. This is the motivation of the study by [28], which analyzes why discourse relation classification is a challenging task and which aspects of discourse relations are difficult for automatic systems.

The proposal contains five dimensions that are rooted in the cognitive approach to coherence relations (CCR) [30, 31]. Some dimensions are added to allow more relations to be distinguished (collectively referred to as "UniDim dimensions" or interchangeably as "dimensions of the UniDim proposal" in the following). An overview of these dimensions is presented below.

Two segments that may stand in a discourse relation are identified first, the two segments being denoted as S_1 and S_2 in their surface order in discourse. The underlying propositions of the two segments are denoted as P and Q, respectively.

One dimension is *basic operation*, which has two values: *causal* and *additive*. A causal relation means that the two segments are strongly connected, and typically, an implication relation $P \rightarrow Q$ can be deduced. In (1), S_2 shows the cause and S_1 gives the consequent. If the two segments are just loosely connected and only a conjunction relation $P \wedge Q$ can be inferred, the value at this dimension is additive, as shown in (2).

(1) [He immigrated to the USA,]$_{S_1}$ because [his natural parents were believed to live there.]$_{S_2}$

(2) [She is a painter]$_{S_1}$ and [her studio is a few blocks away.]$_{S_2}$

As indicated in [23], basic operation can be used to distinguish causal relations and conditional relations from additive or temporal relations.

The second dimension is *source of coherence*. It has two values: semantic and pragmatic in the CCR framework [30], later renamed as *objective* and *subjective* in [32]. A relation is objective if the segments are connected because of their propositional contents, and the relation holds because the connection is coherent based on world knowledge, as shown by (3). A relation is subjective if the speaker's reasoning or the pragmatic effect of the relation is prominent. (4) shows a claim in S_2, and S_1 is a supporting argument.

(3) [It was dark outside,]$_{S_1}$ so [he lit up a candle.]$_{S_2}$

(4) [Smoking is an unhealthy habit]$_{S_1}$ and [we should ban it.]$_{S_2}$

This dimension can be used to distinguish relations that are related to real-world situations, such as temporal sequence and cause-consequence, from argumentative relations, such as claim-argument or evidence-justification [23].

The third dimension is *implication order*. This dimension is only applicable to causal relations, which can be categorized into relations with basic orders and non-basic orders of implication. As additive relations are generally symmetric, this dimension is not applicable. For a causal relation characterized by $P \rightarrow Q$, if S_1 expresses P and S_2 expresses Q (note that S_1 and S_2 are in linear order), then this relation is in basic order, as shown in (5). If S_2 actually expresses P, while S_1 expresses Q, this relation is in non-basic order, as shown in (6).

(5) [He received a warning message,]$_{S_1}$ [so he did not attend the conference.]$_{S_2}$

(6) [He did not attend the conference,]$_{S_1}$ because [he received a message telling him not to go.]$_{S_2}$

It is clear to see that the implication order dimension is mainly used to distinguish relations with directionality, such as cause-result and cause-reason.

The fourth dimension is *polarity*. A relation is characterized by *positive* polarity if the propositions P and Q, expressed by S_1 and S_2, respectively, have the same logical polarity and support each other, as shown in (7). A relation is of *negative* polarity if the relation involves the juxtaposition of $\neg P$ and P or $\neg Q$ and Q in the two segments, as shown in (8). In this example, a positive polarity would require a reason or result that supports the decision of closing the library.

(7) [We like the garden]$_{S_1}$ because [it is pretty.]$_{S_2}$

(8) [The university library was closed]$_{S_1}$ although [students wanted more space for study.]$_{S_2}$

This dimension is useful for characterizing contrastive, adversative, and concession relations [23].

The fifth dimension is *temporality*, which distinguishes between *temporal* and *nontemporal* relations. Under temporal relations, temporality has three values: *synchronous*, *chronological*, and *anti-chronological*. Synchronous relations are temporal relations featuring simultaneous occurrence of events. If events described in the segments happen in temporal order, then the relation is chronological; otherwise, the relation is anti-chronological.

In order to characterize more relations, additional dimensions are introduced, including *specificity*, *lists*, and *alternatives* for additive relations, and *conditionals* and *goal-oriented relations* for causal relations (denoted collectively as "additional dimensions" in the following).

Sanders et al. [23] and other researchers such as [33] try to test if CCR annotations are consistently categorized into relations in the other frameworks. Their investigation reveals that discrepancies between frameworks arise from variations in how coherence relations are defined, the methods used to perform the annotation, and the rules governing discourse segmentation. Consequently, the alignment between discourse relations is generally many to many across different frameworks.

3 Methodology

This section describes the experiments conducted to test how the UniDim proposal can be applied to discourse relation classification across different frameworks. For brevity, we follow the abbreviation convention adopted in [23] to denote the values of UniDim dimensions: *pos* = positive, *neg* = negative, *add* = additive, *cau* = causal, *bas* = basic order of implementation, *non-b* = non-basic order of implementation, *obj* = objective, *sub* = subjective, *chron* = chronological, *anti* = anti-chronological, *syn* = synchronous, *NA* = not applicable, *spec* +: marked by specificity, *spec* –: other cases in specificity, *con* +: marked by conditional, *con* –: other cases in conditional, *goal* +: marked by goal orientedness, and *goal* –: other cases in goal orientedness. The nine UniDim dimensions are abbreviated as *pol* (polarity), *bop* (basic operation), *imp* (implication order), *soc* (source of coherence), *temp* (temporality), *spec* (specificity), *alt* (alternative), *con* (conditional), and *goal* (goal orientedness). As PDTB implicit relation classification is more challenging [34], the current study focuses on PDTB implicit relation classification, in agreement with the mainstream research on shallow discourse parsing.

3.1 Labeling Data with UniDim Dimensions

As UniDim dimensions are not annotated in RST-DT, PDTB 3.0, or other commonly used discourse corpora, a rule-based method is used to obtain their labels. Given an input X with its relation label t, the UniDim dimension labels can be obtained with Appendices A and C in [23] for SDRT and RST, respectively, and Table 10 in [29] for PDTB 3.0 (Appendix B in [23] focuses on PDTB 2.0).

For instance, for RST, if the end label of X is *cause* and the nuclearity is *NS* (the first argument functions as nucleus and the second argument serves as satellite), it can be seen from Appendix C in [23] that the dimension values are *pos*, *cau*, *bas*, *obj*, *chron*, and under-specified for *additional*. Appendices A and C in [23] are attached as Appendix 1 and Appendix 2, and Table 10 in [29] is attached as Appendix 3.

3.2 Upper Limits of Performance

As previously discussed, the alignment between discourse relations is many to many. Moreover, data imbalance poses a challenge, since the distributions of both discourse relations and UniDim dimensions are heavily skewed (see Sect. 4.1). Therefore, we first test how high the performance can be, using gold labels of UniDim dimensions for discourse relation classification.

The set of UniDim dimensions, referred collectively as d_s, is passed to embedding layers before being fed to a multilayer perceptron (MLP) for discourse relation classification. There may be multiple ways to combine the embeddings of

UniDim dimensions. They can be simply concatenated or combined using an attention mechanism. Alternatively, an ensemble model can be used to combine results using the concatenation of the embeddings of UniDim dimensions and using the attention-weighted combination of the embeddings. Empirical results indicate that the attention-weighted combination of the embeddings yields higher performance. The self-attention mechanism proposed by [35] is used in all the experiments.

$$f_{in} = \delta_{att}(E_{pol}, E_{bop}, ..., E_{goal}). \tag{1}$$

Given the input formed by the embeddings of UniDim dimensions, the MLP is trained for discourse relation classification.

$$f_{cls} = \theta_{MLP} \cdot (f_{in}). \tag{2}$$

The training objective is to minimize the loss of discourse relation classification only. The canonical cross-entropy loss is used:

$$\mathcal{L}_r = -\sum_{i=1}^{N}\sum_{l=1}^{C} c_l^i \log p(c_l^i), \tag{3}$$

where N is the batch size, C is the total number of classes, and $p(c_l^i)$ is the probability predicted for a class c.

Note that the UniDim dimensions are provided for the end labels of the relation sets used in the corpora under different frameworks, while the target relation labels are the 16 broad classes for RST, PDTB L2 senses, and the classes in SDRT. In this sense, the mapping between the SDRT labels and UniDim dimensions is unambiguous, since a target label is mapped consistently to a fixed set of UniDim dimensions. However, even in this case, it is anticipated that the performance will not be 100% accurate because of imbalanced data distribution of the target labels. The classification performance is taken as an upper limit of using UniDim dimensions for discourse relation classification in different frameworks.

3.3 Discourse Relation Classification Using UniDim Dimensions

This set of experiments is conducted to check the performance attainable using predicted UniDim dimensions for discourse relation classification. Figure 2 shows a schematic diagram of the model for the experiments, forming a stacking classifier. The input X is fed to the classifiers of UniDim dimensions, and the predicted UniDim dimensions are fed to embedding layers. A classification head of discourse relations is configured on top of the attention-weighted combination of the embeddings. The training objective is to minimize both the loss of discourse relation classification and the classification losses of the nine UniDim dimensions. A simple addition of the

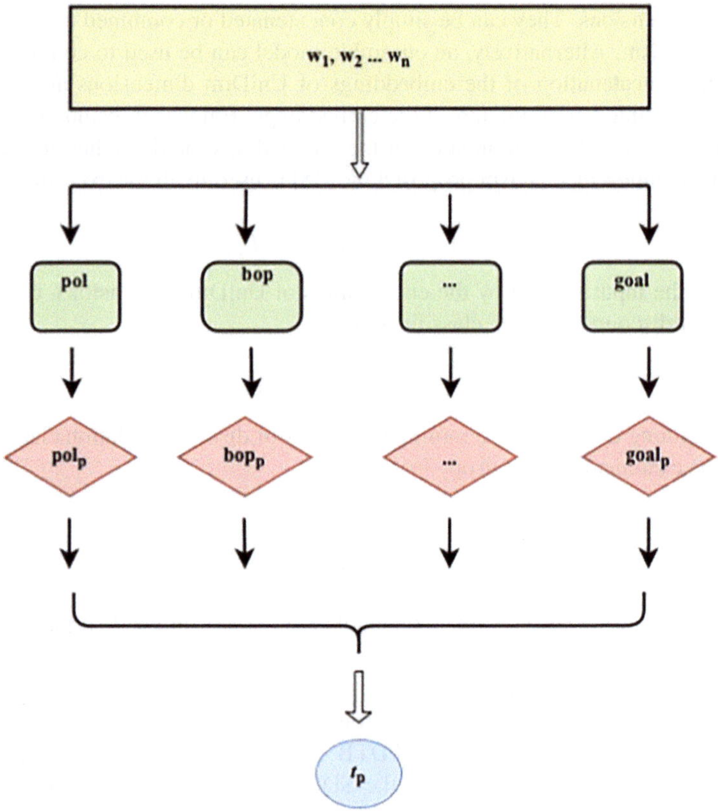

Fig. 2 A diagram shows the model using predicted UniDim dimensions for discourse relation classification, which takes a sequence of input, $w_1, w_2, \ldots, w_n$, and predicts UniDim dimensions pol, bop, $\ldots$, $goal$. The embeddings of predicted UniDim dimensions are combined for predicting discourse relations. The diamonds represent embeddings of the predicted UniDim dimensions. The bottom circle in blue shows the final objective of discourse relation classification

losses is adopted:

$$\mathcal{L}_{total} = \mathcal{L}_r + \mathcal{L}_{pol} + \mathcal{L}_{bop} + \ldots + \mathcal{L}_{goal} \tag{4}$$

where L_r denotes the loss for discourse relation classification, and L_{pol}, L_{bop}, and so on represent the losses for the classification of the nine UniDim dimensions.

Compared with the experiments shown above, this set of experiments tests the performance of only relying on UniDim dimensions for discourse relation classification, which reflects the performance using UniDim dimensions in a realistic setting.

As can be understood from the discussion in Sect. 2, the UniDim dimensions are not independent. For instance, the implication order dimension is only applicable to causal relations, a category under the basic operation dimension. Therefore, the results for the classification of the basic operation dimension influence the value of the implication order dimension. Inspired by the method proposed by [36], which captures label dependencies by conditioning the prediction of child labels on their parents, the embeddings of the predicted parent dimension are passed as a part of input for the classification of a child dimension. In this sense, the classification result of the basic operation dimension is passed as a part of input for the classification of *alt*, *spec*, *con*, *goal*, *temp*, and *imp*.

4 Experiments

4.1 Data

As this study focuses on performance under different discourse frameworks, the PDTB 3.0 (eng.pdtb.pdtb) and STAC (eng.sdrt.stac) data from the DISRPT benchmark [18] are extracted. Based on the method described in Sect. 3.1, to obtain labels of UniDim dimensions, the relation labels are sufficient for SDRT, and the end labels and the surface order of argument 1 and argument 2 are needed for PDTB, where the surface order information is available from DISRPT. For RST, the end labels and nuclearity information are required. However, DISRPT does not include annotations of nuclearity. Therefore, we use the method described in [37] to obtain RST data, which involves converting RST trees into dependency structure and extracting pairs of EDUs, thus keeping the input data in DISRPT but making nuclearity information available.

Following [38], data of less frequent relations (occurrence $<$100) are filtered in PDTB implicit relation classification, and 14-way classification at L2 is adopted.

In [23], the corpus created by [39] is used for analyzing SDRT. Therefore, the results for the mapping between SDRT relations and UniDim dimensions are limited to the 12 relations used by [39], but some of these relations do not appear in STAC. Since the present study does not focus on identifying the mapping between discourse relations and UniDim dimensions, relations in STAC that are not covered by the analysis in [23] are filtered. However, in future work, with an analysis of the relationship between discourse relations and UniDim dimensions, this part of data can be included.

In some cases, the value for a UniDim dimension is ambiguous, for instance, *alternation* in SDRT is assigned a label of *obj/sub* at *soc*. Following the approach adopted by [28] and [29], these values are categorized as "NS." The number of classes under each UniDim dimension after preprocessing is shown in Table 2, which is the same for the three frameworks.

Figure 3 shows the distributions of UniDim dimensions. It is clear that all the distributions are heavily skewed. Figures 4, 5, and 6 show the relation label distributions for the training sets of RST, PDTB implicit relations, and SDRT, respectively. Given the skewed distributions of the UniDim dimensions and target relation labels, we can conclude that the task presents significant challenges.

Table 2 Number of classes under each UniDim dimension

	NO. of classes
pol	3 (NS, pos, neg)
bop	3 (NS, add, cau)
imp	4 (NS, NA, bas, non-b)
soc	3 (NS, obj, sub)
temp	4 (NS, anti, chron, sync)
spec	2 (–, +)
alt	2 (–, +)
con	2 (–, +)
goal	2 (–, +)

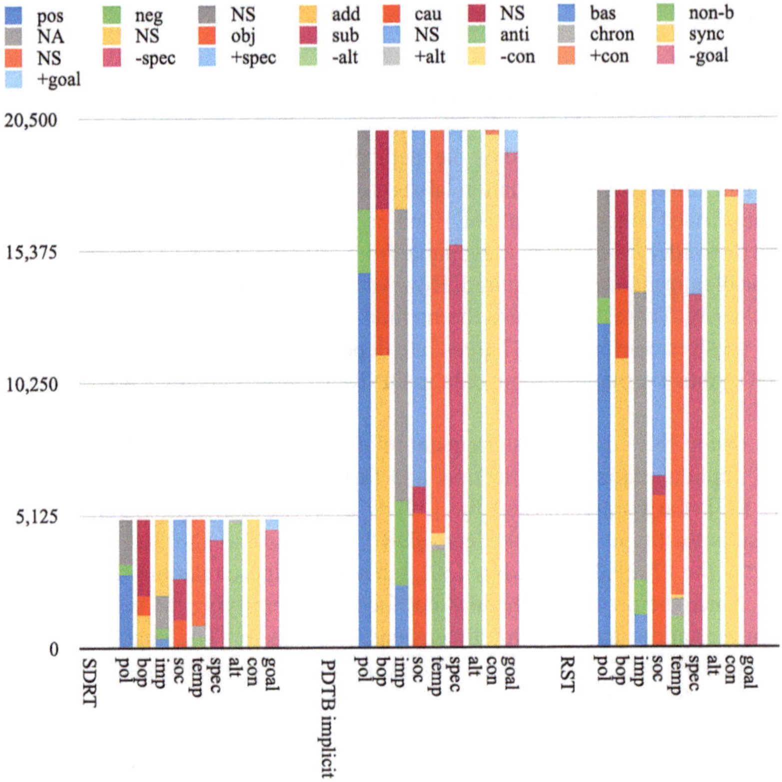

Fig. 3 Statistics of UniDim dimensions. The caption labels can be read from left to right following the order in the bar chart. The y-axis shows the frequency of each of the UniDim dimensions in STAC, PDTB implicit relation data, and RST-DT. The bar chart highlights the imbalanced distributions of UniDim dimensions across different frameworks

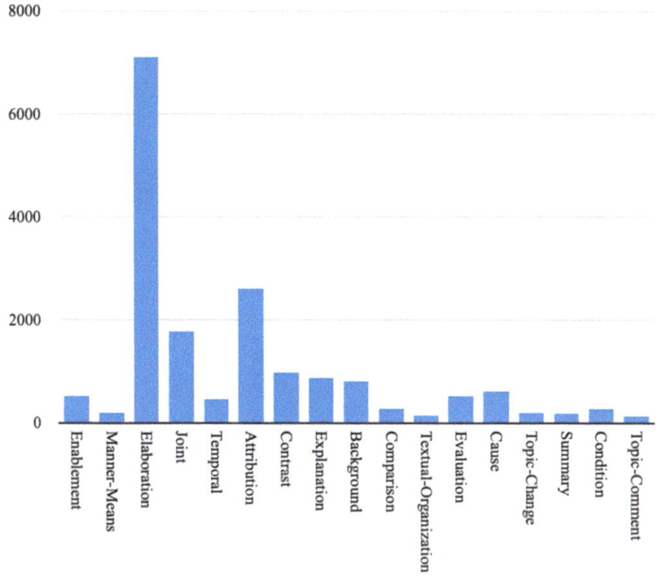

Fig. 4 Distribution of RST relations. It can be seen that *elaboration* has the highest frequency, significantly surpassing the others. Other notable categories include *joint* and *attribution*

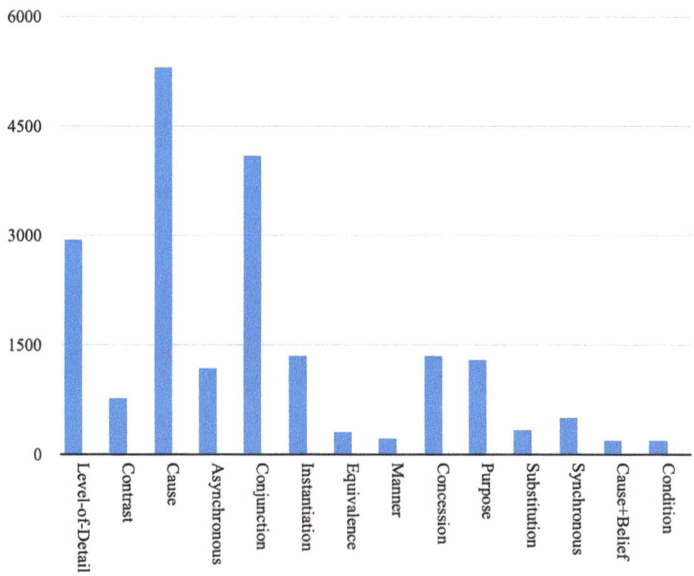

Fig. 5 Distribution of PDTB implicit relations. *Cause* and *conjunction* are the most frequent relations, while *condition, cause+belief* and *manner* are significantly less common

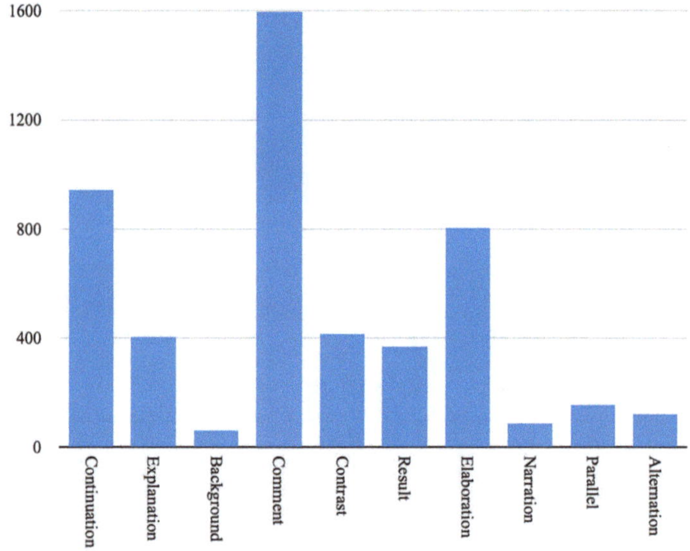

Fig. 6 Distribution of SDRT relations. The *comment* relation is the most frequent in STAC while *background* and *narration* are the least frequent

4.2 Experimental Configurations

The *roberta-base* model from the Transformers library [40] is used as the input encoder. Thus, the baseline method is fine-tuning the *roberta-base* model for discourse relation classification, involving no UniDim dimensions.

The maximum sequence length is set to 256, and the learning rate is $1e$-5 with a warm-up ratio of 0.06. The gradient clipping value is set to 2.0. L2 regularization is adopted. The random seed is 106524.

The maximum number of epochs is set to 30. An early-stop mechanism is employed, with a patience value of seven epochs monitoring the performance on the validation set. The embedding dimension size for UniDim dimensions is 50 * number of classes, and the activation function of MLP is ReLU.

The experiments are implemented with the PyTorch framework [41], and a single RTX A5000 GPU is used for all the experiments.

4.3 Experimental Results

In this section, the experimental results are reported. Unless stated otherwise, accuracy and the macro-averaged F1 score are used.

4.3.1 Baseline Performance and Upper-Limit Performance

Results of discourse relation classification with the baseline method and the upper limits using gold UniDim dimensions are shown in Table 3. It is clear that results that

Table 3 Baseline (B.) and upper-limit (U.) results (%)

	B. Acc.	B. F1	U. Acc.	U. F1
RST	63.38	43.89	83.40	58.74
PDTB implicit	61.35	57.32	89.04	75.71
SDRT	56.23	46.24	98.25	88.37

can be obtained with gold UniDim dimensions are very high. With DISRPT 2023 [42], the highest performance for RST-DT is 65.71 in accuracy, and for STAC, it is 60.79. As the present study focuses on PDTB implicit relations and the preprocessing step is different from that in [42], the results for PDTB are not directly comparable. The baseline method, simple as it is, achieves strong performance. As has been indicated in Sect. 3.2, the mapping between SDRT relations and UniDim dimensions is unambiguous, and therefore, the performance is near perfect. However, due to imbalanced distribution of the target labels, the performance shown by F1 score is not as high as that demonstrated by accuracy score.

4.3.2 Results of Using Predicted UniDim Dimensions for Discourse Relation Classification

The results of the set of experiments described in Sect. 3.3 are shown in Table 4. It can be seen that using the stacking classifier causes large performance drops for all the frameworks compared with the baselines. While the UniDim dimensions are conceptually simple and the upper-limit performance shows promising results, only using predicted UniDim dimensions does not seem to be sufficient for discourse relation classification.

Results on Classification of UniDim Dimensions The performance on the classification of UniDim dimensions is shown in Table 5. Owing to skewed distributions of UniDim dimensions, F1 scores are shown. As shown in Table 2, UniDim dimensions have different numbers of classes. Since the distributions are heavily imbalanced, majority class guessing is adopted to compute the baselines, which means using the majority class as the predicted value for all the test data. The F1 scores shown in Table 5 indicate that the performance on the classification of UniDim dimensions is fairly high. However, even with reasonable performance on classifying the UniDim dimensions, the results on discourse relation classification show considerable degradation compared with the baseline. No correlations are observed between the results for discourse relation classification and the results for the classification of

Table 4 Results of the stacking classifier (%), compared with the baseline

	Acc.	F1
RST	57.37 ($\downarrow$ 6.01)	30.45 ($\downarrow$ 13.44)
PDTB implicit	55.18 ($\downarrow$ 6.17)	40.81 ($\downarrow$ 16.51)
SDRT	54.61 ($\downarrow$ 1.62)	42.05 ($\downarrow$ 4.19)

Table 5 Results of the classification of UniDim dimensions (F1). The baselines shown by ".B" are computed based on majority class guessing

	pol	bop	imp	soc	temp	spec	alt	con	goal
RST	75.40	73.08	66.25	56.94	56.48	80.10	83.29	90.72	81.90
RST.B	27.61	25.77	19.33	25.69	23.51	43.53	49.97	49.61	49.22
PDTB	72.86	71.10	67.30	65.86	58.65	80.77	100.00	85.16	79.38
PDTB.B	28.00	24.04	18.03	27.19	21.90	43.76	100.00	49.77	48.88
SDRT	74.19	58.48	51.78	62.11	54.50	72.82	83.05	100.00	65.16
SDRT.B	24.20	24.80	18.60	21.03	30.17	45.59	49.38	100.00	47.88

UniDim dimensions. Moreover, results on the classification of UniDim dimensions differ across the frameworks, with rather mixed results. For example, the results on *bop* and *imp* are much higher for RST and PDTB than for SDRT, while the results on *soc* are significantly lower for RST than for PDTB and SDRT. In combination with Fig. 3, it can be seen that for *bop*, the part in magenta, which means "NS" (under-specified), accounts for a larger proportion in SDRT than PDTB and RST. Similarly, for *imp*, the part in yellow ("NS") is larger in SDRT than in PDTB and RST. For *soc*, the proportion of "NS" (in blue) is higher in RST and PDTB than in SDRT. As the amount of PDTB data is much larger than RST, the results may be higher for PDTB. Therefore, the class "NS" may be the cause for the patterns identified, and it is challenging for automatic prediction. The case with *goal* is different. While the three frameworks show similar distributions, the highest performance is achieved in RST, followed closely by PDTB, with SDRT lagging far behind. Examination of the mapping rules in Appendix 1 shows that in SDRT, only the relation *explanation* is tentatively marked by *goal*. The results may suggest that more efforts are needed to refine the mapping rules so that discourse relations can be mapped correctly to UniDim dimensions and the amount of "NS" cases can be reduced. Moreover, a layer of annotation with UniDim dimensions is likely to be helpful in improving the results of automatic classifiers.

Relation-Level Analysis for RST The relation-level performance for RST is shown in Table 6. It can be seen that six relations including *comparison*, *joint*, *manner-means*, *summary*, *textual-organization*, and *topic-change* have F1 scores of zero or close to zero if predicted UniDim dimensions are used. This is likely to be the reason for the large drop in the overall performance, as the performance on some of the other relations (*elaboration*, *explanation*, and *temporal*) shows a decrease, while there are also relations (*background*, *contrast*, and *topic-comment*) with increased F1 scores. In combination with Fig. 4, it can be seen that except for *joint*, the five relations having F1 scores of zero are classes with a small amount of training data. The confusion matrix (Appendix 4) shows that most of the six relations are predicted as *elaboration*, which is the largest class in RST-DT. This may explain why the precision score of *elaboration* goes down (from 66.08 to 54.48), while the recall score goes up (from 75.38 to 81.78). This finding shows that using predicted UniDim dimensions as intermediate representations exacerbates data imbalance in discourse relation classification.

Table 6 Relation-level performance for RST (P = precision, R = recall). Performance for the baseline is indicated by the subscript *B*. This convention is also used in Table 7 (PDTB) and Table 8 (SDRT)

	P	R	$F1$	P_B	R_B	$F1_B$
Attribution	94.08	96.36	95.21	93.59	97.27	95.39
Background	45.00	40.91	42.86	38.10	36.36	37.21
Cause	22.45	13.75	17.05	22.22	15.00	17.91
Comparison	0.00	0.00	0.00	66.67	41.38	51.06
Condition	87.80	75.00	80.90	90.24	77.08	83.15
Contrast	54.62	48.97	51.64	42.37	51.72	46.58
Elaboration	54.48	81.78	65.39	66.08	75.38	70.42
Enablement	61.54	69.57	65.31	61.40	76.09	67.96
Evaluation	21.88	17.50	19.44	28.26	16.25	20.63
Explanation	31.13	30.00	30.56	41.98	30.91	35.60
Joint	100.00	1.42	2.79	58.04	69.81	63.38
Manner-means	0.00	0.00	0.00	55.56	37.04	44.44
Summary	0.00	0.00	0.00	85.71	37.50	52.17
Temporal	31.65	34.72	33.11	54.05	27.78	36.70
Textual-organization	0.00	0.00	0.00	25.00	22.22	23.53
Topic-change	0.00	0.00	0.00	0.00	0.00	0.00
Topic-comment	33.33	8.33	13.33	0.00	0.00	0.00

Relation-Level Analysis for PDTB The relation-level performance for PDTB is shown in Table 7. Similar to RST, it can be seen that four relations including *cause+belief*, *equivalence*, *instantiation*, and *manner* have F1 scores of zero. As shown in Fig. 5, except for *instantiation*, these relations are classes with a small number of training data. By checking the confusion matrix (Appendix 5), it is found that most of the instances of *cause+belief* are predicted as *cause*, which is unsurprising, since *cause* is a large class and the two relations are similar in representations with UniDim dimensions except for the *soc* dimension. Instances of *equivalence* are predicted as *cause* (29.17%), *conjunction* (25.00%), and *level-of-detail* (29.17%). The representations with UniDim dimensions are similar for *equivalence*, *conjunction*, and *level-of-detail*. Moreover, as *cause* is a large class, while *equivalence* has a small amount of training data, in multi-class classification, it is not rare that the classifier assigns the label of the largest class to classes with a small data amount. For *instantiation*, the confusion matrix indicates that most of the instances are predicted as *level-of-detail*. The two classes have the same representations with UniDim dimensions. Similarly, all the instances of *manner* are predicted as *level-of-detail*, which has the same representation in UniDim dimensions. This may suggest that classes under the L1 sense *expansion* need more fine-grained mapping rules, which echoes the insight shown in [33].

Only *contrast* and *concession* have higher performance than the baseline. As indicated in [23] and [20], it is more challenging to map PDTB implicit relations to the other frameworks as greater ambiguity is involved.

Table 7 Relation-level performance for PDTB

	P	R	$F1$	P_B	R_B	$F1_B$
Concession	64.79	55.42	59.74	71.19	50.60	59.15
Contrast	73.68	53.85	62.22	62.16	44.23	51.69
Cause	63.42	60.15	61.74	67.20	61.99	64.49
Cause+Belief	0.00	0.00	0.00	5.88	10.00	7.41
Condition	85.71	60.00	70.59	77.78	70.00	73.68
Purpose	67.86	41.30	51.35	89.80	95.65	92.63
Conjunction	48.12	79.50	59.95	58.79	66.46	62.39
Equivalence	0.00	0.00	0.00	35.00	29.17	31.82
Instantiation	0.00	0.00	0.00	54.74	74.29	63.03
Level-of-detail	46.56	62.89	53.51	58.42	57.22	57.81
Manner	0.00	0.00	0.00	44.44	100.00	61.54
Substitution	70.00	77.78	73.68	81.48	81.48	81.48
Asynchronous	48.57	51.52	50.00	52.63	60.61	56.34
Synchronous	44.44	21.05	28.57	36.36	42.11	39.02

Table 8 Relation-level performance for SDRT

	P	R	$F1$	P_B	R_B	$F1_B$
Alternation	76.92	71.43	74.07	84.62	78.57	81.48
Background	0.00	0.00	0.00	0.00	0.00	0.00
Comment	63.14	71.49	67.05	64.53	70.66	67.46
Continuation	51.76	56.05	53.82	51.35	60.51	55.56
Contrast	50.75	47.89	49.28	58.57	57.75	58.16
Elaboration	58.54	48.00	52.75	58.26	44.67	50.57
Explanation	38.33	31.94	34.85	28.89	36.11	32.10
Narration	0.00	0.00	0.00	0.00	0.00	0.00
Parallel	43.40	63.89	51.69	80.77	58.33	67.74
Result	35.71	38.46	37.04	50.00	48.72	49.35

Relation-Level Analysis for SDRT The relation-level performance for SDRT is shown in Table 8. Different from RST and PDTB where large drops compared with the baseline method are observed, using predicted UniDim dimensions causes relatively small performance drops for SDRT. Relations including *background* and *narration* have F1 scores of zero in both the baseline and the stacking classifier. As can be seen from Fig. 6, the two relations are classes with the smallest data amounts. The performance on *elaboration* and *explanation* shows slight improvement with the stacking classifier over the baseline, while performance on *alternation*, *contrast*, *parallel*, and *result* shows considerable decreases. Figure 6 shows that *alternation* has a small amount of training data, only slightly higher than *background* and *narration*, and the amount of training data for *parallel* is slightly higher than *alternation*. With the confusion matrix (Appendix 6), it can be seen that a small proportion of instances of *alternation* is classified as *continuation*, but the bulk of the class has correct predictions. The case with *contrast* is similar, but a large number of instances are classified as *comment*, which is the largest class. Similar patterns can be found

Table 9 Results of the correlation strength between discourse relation classification and the classification of UniDim dimensions

	pol	bop	imp	soc	temp	spec	alt	con	goal
RST	45.66	49.89	50.76	45.03	39.12	18.27	2.19	9.22	12.85
PDTB	44.57	54.19	57.45	42.40	34.20	46.14	0.00	5.60	15.08
SDRT	34.05	73.75	74.85	79.39	47.13	43.67	7.31	0.00	37.22

for *parallel* and *result*, where the majority of the instances are classified correctly, but the incorrect predictions are spread over a variety of classes, each taking up a small proportion. This suggests high uncertainty in the predictions.

Quantitative Analysis of Contribution of Different UniDim Dimensions on Discourse Relation Classification The UniDim dimensions have different strengths in characterizing discourse relations. For instance, additional dimensions including *spec*, *alt*, *con*, and *goal* are only applicable to some relations. Hence, correlation test is performed to check which of the nine UniDim dimensions are important for the results on discourse relation classification for the three frameworks. Cramér's V [43] is used, which measures the correlation strength (in a range of [0, 1]) between two categorical variables. Table 9 shows the results. For RST, *pol*, *bop*, *imp*, *soc*, and *temp* are UniDim dimensions that have major influence on the results of discourse relation classification. For PDTB, in addition to the five UniDim dimensions, *spec* has considerable influence, which is expected, since some relations under the L1 sense *Expansion* have a large amount of training data, such as *level-of-detail* and *instantiation*, and these relations are marked by this dimension. For SDRT, apart from *pol*, *bop*, *imp*, *soc*, *temp*, and *spec*, the performance on *goal* is strongly connected with the results on discourse relation classification. Based on the mapping rules between discourse relations and UniDim dimensions, only the relation *explanation* is marked by this dimension. As shown in Fig. 6, *Explanation* is a class with a relatively large amount of training data. This may explain why the *goal* dimension also has a large influence on the results.

5 Conclusion and Future Work

We show how the UniDim proposal by [23] can be used for discourse relation classification across different frameworks, such as RST, PDTB, and SDRT, on the DISRPT benchmark [18], where discourse segmentation and relation taxonomies differ considerably. With cognitively inspired primitive dimensions, discourse relations, regardless of the framework, can be decomposed and expressed by the interlingua of UniDim dimensions. The upper limits attainable are quite high, and reasonable performance can be achieved on the automatic classification of the UniDim dimensions. However, the performance achieved only using predicted UniDim dimensions for discourse relation classification is much lower than the baseline. We find that with this approach, the performance on classes with a small amount of training data drops considerably. Moreover, under-specified classes under UniDim dimensions are challenging for automatic classifiers. This calls for more efforts on developing fine-grained mapping rules between discourse relations and UniDim dimensions, and annotations of UniDim dimensions in addition to existing RST-style and PDTB-style discourse annotations are likely to be helpful in improving the performance of automatic means. In future work, studies can be performed on how to combine data of different frameworks efficiently to improve the performance on the classification of UniDim dimensions and discourse relations.

Appendix 1: Mapping Between SDRT Relations and UniDim Dimensions

See Fig. 7.

SDRT	Polarity	Basic op.	Impl. order	SoC	Temporality	Add. features
Alternation	pos/neg	add/cau	any	obj/sub	(chron/anti)	(conditional), (alternative)
Background	pos	add	N.A.	obj	(chron/anti/syn)	
Commentary	pos	add	N.A.	sub	N.A.	
Consequence	pos	cau	bas/non-b	obj/sub	chron/anti	conditional
Continuation	pos	add	N.A.	obj/sub	N.A.	list
Contrast	neg	add/cau	any	obj/sub	(chron/anti)	
Elaboration	pos	add/cau	any	obj/sub	(chron/anti)	specificity, (list)
Explanation	pos	cau	non-b	obj	anti	(goal)
Narration	pos	add	N.A.	obj	chron	
Parallel	pos	add	N.A.	obj	N.A.	
Precondition	pos	add	N.A.	obj	anti	
Result	pos	cau	bas	obj	chron	

Fig. 7 Mapping between SDRT relations and UniDim dimensions, from [23]

Appendix 2: Mapping Between RST Relations and UniDim Dimensions

See Table 10.

Table 10 Mapping between RST relations and UniDim dimensions, from [23]

Class	End label	Nuc.	N-S	pol	bop	imp	soc	temp	add.
Background	Background	Mono	N-S	pos/neg	add	N.A.	obj	anti/N.A.	
	Background	Mono	S-N	pos/neg	add	N.A.	obj	chron/N.A.	
	Circumstance	Mono		pos/neg	add	N.A.	obj	syn/N.A.	
Cause	Cause	Mono	N-S	pos	cau	bas	obj	chron	
	Cause	Mono	S-N	pos	cau	non-b	obj	anti	
	Cause-result	Multi		pos	cau	bas/non-b	obj	chron/anti	
	Result	Mono	N-S	pos	cau	non-b	obj	anti	
	Result	Mono	S-N	pos	cau	bas	obj	chron	
	Consequence-n	Mono	N-S	pos	cau	non-b	obj	anti	
	Consequence-n	Mono	S-N	pos	cau	bas	obj	chron	
	Consequence-s	Mono	N-S	pos	cau	bas	obj	chron	
	Consequence-s	Mono	S-N	pos	cau	non-b	obj	anti	
	Consequence	Multi		pos	cau	bas/non-b	obj	chron/anti	
Comparison	Comparison	Both		pos	add	N.A.	obj/sub	N.A.	
	Preference	Mono		neg	add	N.A.	obj/sub	N.A.	
	Analogy	Both		pos	add	N.A.	sub	N.A.	
	Proportion	Multi		pos	add/cau	any	obj/sub	any	
Conditional	Condition	Mono	N-S	pos/neg	cau	non-b	obj/sub	anti/N.A.	conditional
	Condition	Mono	S-N	pos/neg	cau	bas	obj/sub	chron/N.A.	conditional
	Hypothetical	Mono	N-S	pos	cau	non-b	sub	N.A.	conditional
	Hypothetical	Mono	S-N	pos	cau	bas	sub	N.A.	conditional
	Contingency	Mono	N-S	pos/neg	cau	non-b	obj	anti	conditional
	Contingency	Mono	S-N	pos/neg	cau	bas	obj	chron	conditional
	Otherwise	Mono	N-S	neg	cau	bas	obj/sub	chron/N.A.	conditional
	Otherwise	Multi		neg	cau	bas	obj/sub	chron/N.A.	conditional
Contrast	Contrast	Multi		neg	add	N.A.	obj/sub	any	
	Concession	Mono	N-S	neg	cau	non-b	obj/sub	anti/N.A.	
	Concession	Mono	S-N	neg	cau	bas	obj/sub	chron/N.A.	
	Antithesis	Mono		neg	add/cau	any	obj/sub	any	
Elaboration	El.-additional	Mono		pos	add	N.A.	obj/sub	N.A.	
	El.-gen.-spec.	Mono		pos	add	N.A.	obj/sub	N.A.	specificity
	El.-part-whole	Mono		pos	add	N.A.	obj	N.A.	specificity
	El.-process-step	Mono		pos	add	N.A.	obj	N.A.	specificity
	El.-object-attr.	Mono		pos	add	N.A.	obj	N.A.	specificity
	El.-set-member	Mono		pos	add	N.A.	obj	N.A.	spec.-ex.
	Example	Mono		pos	add	N.A.	obj	N.A.	spec.-ex.
	Definition	Mono		pos	add	N.A.	obj	N.A.	specificity
Enablement	Purpose	Mono	N-S	pos	cau	bas	obj/sub	chron/N.A.	goal
	Purpose	Mono	S-N	pos	cau	non-b	obj/sub	anti/N.A.	goal
	Enablement	Mono	N-S	pos	cau	non-b	obj/sub	anti/N.A.	goal
	Enablement	Mono	S-N	pos	cau	bas	obj/sub	chron/N.A.	goal
Evaluation	Evaluation	Both		pos	add/cau	any	sub	N.A.	specificity
	Interpretation	Both		pos	add/cau	any	sub	N.A.	specificity
	Conclusion	Mono	N-S	pos	cau	bas	sub	N.A.	specificity
	Conclusion	Mono	S-N	pos	cau	non-b	sub	N.A.	specificity
	Conclusion	Multi		pos	cau	bas/non-b	sub	N.A.	specificity
	Comment	Mono		pos	add	N.A.	sub	N.A.	specificity
Explanation	Evidence	Mono	N-S	pos	cau	non-b	sub	anti	
	Evidence	Mono	S-N	pos	cau	bas	sub	chron	
	Exp.-argument.	Mono	N-S	pos	cau	non-b	obj	anti	
	Exp.-argument.	Mono	S-N	pos	cau	bas	obj	chron	
	Reason	Mono	N-S	pos	cau	non-b	obj	anti	
	Reason	Mono	S-N	pos	cau	bas	obj	chron	
	Reason	Multi		pos	cau	bas/non-b	obj	chron/anti	
Joint	List	Multi		pos	add	N.A.	obj/sub	syn/chron/N.A.	list
	Disjunction	Multi		pos/neg	add	N.A.	obj/sub	syn/N.A.	alternative
Summary	Summary	Mono		pos	add	N.A.	obj	N.A.	specificity
	Restatement	Mono		pos	add	N.A.	obj	N.A.	spec.-equiv.
Temporal	Temp.-before	Mono	N-S	pos	add	N.A.	obj	chron	
	Temp.-before	Mono	S-N	pos	add	N.A.	obj	anti	
	Temp.-after	Mono	N-S	pos	add	N.A.	obj	anti	
	Temp.-after	Mono	S-N	pos	add	N.A.	obj	chron	
	Temp.-same-time	Both		pos	add	N.A.	obj	syn	
	Sequence	Multi		pos	add	N.A.	obj	chron	
	Inverted-seq.	Multi		pos	add	N.A.	obj	anti	
Manner-means	Means	Mono	N-S	pos	cau	non-b	obj	anti	
	Means	Mono	S-N	pos	cau	bas	obj	chron	goal
Topic-comment	Problem-sol.-n	Mono	N-S	pos	cau	non-b	obj/sub	anti/N.A.	goal
	Problem-sol.-n	Mono	S-N	pos	cau	bas	obj/sub	chron/N.A.	goal
	Problem-sol.-s	Mono	N-S	pos	cau	bas	obj/sub	chron/N.A.	goal
	Problem-sol.-s	Mono	S-N	pos	cau	non-b	obj/sub	anti/N.A.	goal
	Problem-sol.	Multi		pos	cau	bas/non-b	obj/sub	achron/anti/N.A.	goal

Appendix 3: Mapping Between PDTB Relations and UniDim Dimensions

See Table 11.

Table 11 Mapping between PDTB relations (PDTB 3.0) and UniDim dimensions, from [29]

Class_type	End label	A1-A2	pol	bop	imp	soc	temp	add.
Temporal								
Synchronous			pos	add	N.A.	obj	sync	
Asynchronous	Precedence	A1-A2	pos	add	N.A.	obj	chron	
	Precedence	A2-A1	pos	add	N.A.	obj	anti	
	Succession	A1-A2	pos	add	N.A.	obj	anti	
	Succession	A2-A1	pos	add	N.A.	obj	chron	
Contingency								
Cause	Reason	A1-A2	pos	cau	non-b	obj	anti	
	Reason	A2-A1	pos	cau	bas	obj	chron	
	Result	A1-A2	pos	cau	bas	obj	chron	goal
	Result	A1-A2	pos	cau	bas	obj	chron	goal
	NegResult		neg	cau	bas	obj	chron	
Cause+Belief	Reason+Belief	A1-A2	pos	cau	non-b	sub	NS	
	Reason+Belief	A2-A1	pos	cau	bas	sub	NS	
	Result+Belief	A1-A2	pos	cau	bas	sub	NS	
	Result+Belief	A2-A1	pos	cau	non-b	sub	NS	
Cause +SpeechAct	Reason+SpeechAct	A1-A2	pos	cau	non-b	sub	NS	
	Reason+SpeechAct	A2-A1	pos	cau	bas	sub	NS	
	Result+SpeechAct	A1-A2	pos	cau	bas	sub	NS	
	Result+SpeechAct	A2-A1	pos	cau	non-b	sub	NS	
Purpose	arg1-as-goal	A1-A2	pos	cau	non-b	obj/sub	NS	goal
	arg1-as-goal	A2-A1	pos	cau	bas	obj/sub	NS	goal
	arg2-as-goal	A1-A2	pos	cau	bas	sub	NS	goal
Condition	arg1-as-cond	A1-A2	pos	cau	bas	obj/sub	NS	conditional
	arg1-as-cond	A2-A1	pos	cau	non-b	obj/sub	NS	conditional
	arg2-as-cond	A1-A2	pos	cau	non-b	obj/sub	NS	conditional
	arg2-as-cond	A2-A1	pos	cau	bas	obj/sub	NS	conditional
Condition +SpeechAct			pos	cau	bas	sub	NS	conditional
Negative -Condition	arg1-as-negcond	A1-A2	neg	cau	bas	sub	NS	conditional
	arg1-as-negcond	A2-A1	neg	cau	non-b	sub	NS	conditional
	arg2-as-negcond	A1-A2	neg	cau	non-b	sub	NS	conditional
	arg2-as-negcond	A2-A1	neg	cau	bas	sub	NS	conditional
Negative- condition+ speechAct			neg	cau	bas	sub	NS	conditional
Comparison								
Concession	arg1-as-denier	A1-A2	neg	cau	non-b	obj/sub	NS	
	arg1-as-denier	A2-A1	neg	cau	bas	obj/sub	NS	
	arg2-as-denier	A1-A2	neg	cau	bas	obj/sub	NS	
	arg2-as-denier	A2-A1	neg	cau	non-b	obj/sub	NS	
Concession +SpeechAct			neg	cau	bas	sub	NS	
Contrast			neg	add	NA	obj	NS	
Similarity			pos	add	NA	obj	NS	
Expansion								
Conjunction			pos	add	NA	obj/sub	NS	
Disjunction			neg	add	NA	obj/sub	NS	alternative
Equivalence			pos	add	NA	obj/sub	NS	
Exception	arg1-as-excpt		neg	add	NA	obj/sub	NS	
	arg2-as-excpt		neg	add	NA	obj/sub	NS	
Instantiation	arg1-as-instance		pos	add	NA	obj/sub	NS	specificity
	arg2-as-instance		pos	add	NA	obj/sub	NS	specificity
Level-of-detail	arg1-as-detail		pos	add	NA	obj/sub	NS	specificity
	arg2-as-detail		pos	add	NA	obj/sub	NS	specificity
Manner	arg1-as-manner	A1-A2	pos	add	NA	obj/sub	NS	specificity
	arg2-as-manner		pos	add	NA	obj/sub	NS	specificity
Substitution	arg1-as-subst	A1-A2	neg	cau	bas	obj/sub	NS	
	arg1-as-subst	A2-A1	neg	cau	non-b	obj/sub	NS	
	arg2-as-subst	A1-A2	neg	cau	non-b	obj/sub	NS	
	arg2-as-subst	A2-A1	neg	cau	bas	obj/sub	NS	

Appendix 4: Confusion Matrix of RST Relation Classification Results

See Table 12.

Table 12 Confusion matrix of the relation classification results for RST (rows: True; columns: Predicted)

	Attr.	Backgr.	Cau.	Compar.	Cond.	Contr.	Elab.	Enable.	Eval.	Expl.	Joint	M.-M.	Sum.	Tempor.	T-Org.	T-Ch.	T-Comm.
Attr.	318	0	1	0	0	0	9	0	1	1	0	0	0	0	0	0	0
Backgr.	2	45	5	0	3	6	31	0	4	5	0	0	0	8	0	0	1
Cau.	1	2	11	0	0	1	43	3	2	14	0	0	0	3	0	0	0
Compar.	0	1	0	0	0	6	20	1	0	0	0	0	0	1	0	0	0
Cond.	0	3	1	0	36	1	5	0	1	0	0	0	0	1	0	0	0
Contr.	0	1	1	0	1	71	59	1	4	2	0	0	0	5	0	0	0
Elab.	6	16	12	0	1	24	651	11	20	28	0	0	0	25	0	0	2
Enable.	0	0	1	0	0	0	11	32	0	1	0	0	0	1	0	0	0
Eval.	0	4	3	0	0	6	47	0	14	5	0	0	0	1	0	0	0
Expl.	4	6	5	0	0	6	42	2	7	33	0	0	0	4	0	0	1
Joint	0	2	2	0	0	4	187	1	4	6	3	0	0	3	0	0	0
M.-M.	1	2	2	0	0	0	11	0	1	8	0	0	0	2	0	0	0
Sum.	1	0	0	0	0	3	26	0	2	0	0	0	0	0	0	0	0
Tempor.	0	16	3	0	0	1	24	1	0	2	0	0	0	25	0	0	0
T-Org.	2	0	0	0	0	0	7	0	0	0	0	0	0	0	0	0	0
T-Ch.	1	1	0	0	0	0	9	0	2	0	0	0	0	0	0	0	0
T-Comm.	2	1	2	0	0	1	13	0	2	1	0	0	0	0	0	0	2

Appendix 5: Confusion Matrix of PDTB Relation Classification Results

See Table 13.

Table 13 Confusion matrix of the relation classification results for PDTB (rows: True; columns: Predicted)

	Concess.	Contr.	Cau.	Cau.+Bel.	Cond.	Purp.	Conj.	Equiv.	Instan.	Level-of-D.	Manner	Subst.	Asynch.	Synch.
Concess.	46	3	12	0	0	0	16	0	0	4	0	2	0	0
Contr.	5	28	3	0	0	0	9	0	0	4	0	1	2	0
Cau.	6	2	163	0	0	6	41	1	0	41	0	2	8	1
Cau. +Bel.	0	0	8	0	0	0	1	0	0	1	0	0	0	0
Cond.	1	0	1	0	6	1	1	0	0	0	0	0	0	0
Purp.	1	0	24	0	1	19	0	0	0	0	0	1	0	0
Conj.	4	3	10	0	0	0	128	0	0	10	0	1	4	1
Equiv.	2	1	7	0	0	0	6	0	0	7	0	0	1	0
Instan.	0	0	1	0	0	0	4	0	0	64	0	0	1	0
Level-of-D.	5	1	21	0	0	1	40	0	0	122	0	2	2	0
Manner	0	0	0	0	0	0	0	0	0	4	0	0	0	0
Subst.	0	0	2	0	0	0	2	0	0	1	0	21	0	1
Asynch.	1	0	4	0	0	0	6	0	0	3	0	0	17	2
Synch.	0	0	1	0	0	1	12	0	0	1	0	0	0	4

Appendix 6: Confusion Matrix of SDRT Relation Classification Results

See Table 14.

Table 14 Confusion matrix of the relation classification results for SDRT (rows: True; columns: Predicted)

	Alter.	Backgr.	Comm.	Continu.	Contr.	Elab.	Expl.	Narr.	Paral.	Res.
Alter.	10	0	0	2	0	1	0	0	0	1
Backgr.	0	0	5	2	2	2	2	0	0	1
Comm.	0	0	173	25	10	11	12	0	5	6
Continu.	0	0	25	88	5	14	9	0	11	5
Contr.	0	0	15	9	34	6	4	0	1	2
Elab.	2	0	21	26	6	72	7	0	8	8
Expl.	0	0	22	7	6	8	23	0	4	2
Narr.	0	0	0	3	0	2	0	0	0	2
Paral.	1	0	5	3	3	1	0	0	23	0
Res.	0	0	8	5	1	6	3	0	1	15

References

1. Ji, H., & Huang, M. (2021). DiscoDVT: Generating long text with discourse-aware discrete variational transformer. In *Proceedings of the 2021 Conference on Empirical Methods in Natural Language Processing* (pp. 4208–4224).
2. Smith, K. S. (2017). On integrating discourse in machine translation. In *Proceedings of the Third Workshop on Discourse in Machine Translation* (pp. 110–121).
3. Prasad, R., Miltsakaki, E., Dinesh, N., Lee, A., Joshi, A., Robaldo, L., & Webber, B. (2007). The Penn Discourse Treebank 2.0 Annotation Manual. December, 17, 2007.
4. Atwell, K., Li, J. J., & Alikhani, M. (2021). Where are we in discourse relation recognition? In *Proceedings of the 22nd Annual Meeting of the Special Interest Group on Discourse and Dialogue* (pp. 314–325).
5. Yung, F., Ahmad, M., Scholman, M., & Demberg, V. (2024). Prompting implicit discourse relation annotation. In *Proceedings of the 18th Linguistic Annotation Workshop (LAW-XVIII)* (pp. 150–165).
6. Zeldes, A., Liu, Y. J., Iruskieta, M., Muller, P., Braud, C., & Badene, S. (2021). The DISRPT 2021 shared task on elementary discourse unit segmentation, connective detection, and relation classification. In *Proceedings of the 2nd Shared Task on Discourse Relation Parsing and Treebanking (DISRPT 2021)* (pp. 1–12).
7. Fu, Y. (2022). Towards unification of discourse annotation frameworks. In *Proceedings of the 60th Annual Meeting of the Association for Computational Linguistics: Student Research Workshop* (pp. 132–142).
8. Mann, W., & Thompson, S. (1988). Rhetorical structure theory: Toward a functional theory of text organization. *Text -Interdisciplinary Journal for the Study of Discourse, 8*(3), 243–281. https://doi.org/10.1515/text.1.1988.8.3.243
9. Asher, N., & Lascarides, A. (2003). *Logics of conversation*. Cambridge University Press.

10. Webber, B., Stone, M., Joshi, A., & Knott, A. (2003). Anaphora and discourse structure. *Computational Linguistics, 29*(4), 545–587.
11. Webber, B. (2004). D-LTAG: Extending lexicalized TAG to discourse. *Cognitive Science, 28*(5), 751–779.
12. Carlson, L., Marcu, D., & Okurowski, M. E. (2003). Building a discourse-tagged corpus in the framework of rhetorical structure theory. *Current and New Directions in Discourse and Dialogue*, 85–112.
13. Zeldes, A. (2017). The GUM corpus: Creating multilayer resources in the classroom. *Language Resources and Evaluation, 51*(3), 581–612.
14. Afantenos, S., Asher, N., Benamara, F., Bras, M., Fabre, C., Ho-Dac, L. M., ... & Vieu, L. (2012). An empirical resource for discovering cognitive principles of discourse organisation: The ANNODIS corpus. In *Eight International Conference on Language Resources and Evaluation (LREC 2012)* (pp. 2727–2734). European Language Resources Association (ELRA).
15. Asher, N., Hunter, J., Morey, M., Benamara, F., & Afantenos, S. (2016). Discourse structure and dialogue acts in multiparty dialogue: The STAC corpus. In *10th International Conference on Language Resources and Evaluation (LREC 2016)* (pp. 2721–2727).
16. Webber, B., Prasad, R., Lee, A., & Joshi, A. (2019). The Penn Discourse Treebank 3.0 Annotation Manual. Philadelphia, University of Pennsylvania, 35, 108.
17. Zeldes, A., Das, D., Maziero, E. G., Antonio, J., & Iruskieta, M. (2019). The DISRPT 2019 shared task on elementary discourse unit segmentation and connective detection. In *Proceedings of the Workshop on Discourse Relation Parsing and Treebanking 2019* (pp. 97–104).
18. Braud, C., Zeldes, A., Rivière, L., Liu, Y. J., Muller, P., Sileo, D., & Aoyama, T. (2024). DISRPT: A multilingual, multi-domain, cross-framework benchmark for discourse processing. In *Proceedings of the 2024 Joint International Conference on Computational Linguistics, Language Resources and Evaluation (LREC-COLING 2024)* (pp. 4990–5005).
19. Carlson, L., & Marcu, D. (2001). Discourse tagging reference manual. *ISI Technical Report ISI-TR-545, 54*(2001), 56.
20. Demberg, V., Scholman, M. C., & Asr, F. T. (2019). How compatible are our discourse annotation frameworks? Insights from mapping RST-DT and PDTB annotations. *Dialogue & Discourse, 10*(1), 87–135.
21. Poláková, L., Mírovský, J., & Synková, P. (2017). Signalling implicit relations: A PDTB-RST comparison. *Dialogue & Discourse, 8*(2), 225–248.
22. Fu, Y. (2024). Automatic alignment of discourse relations of different discourse annotation frameworks. In *Proceedings of the 20th Joint ACL-ISO Workshop on Interoperable Semantic Annotation@ LREC-COLING 2024* (pp. 27–38).
23. Sanders, T. J., Demberg, V., Hoek, J., Scholman, M. C., Asr, F. T., Zufferey, S., & Evers-Vermeul, J. (2021). Unifying dimensions in coherence relations: How various annotation frameworks are related. *Corpus Linguistics and Linguistic Theory, 17*(1), 1–71.
24. Hovy, E. H., & Maier, E. (1992). Parsimonious or profligate: How many and which discourse structure relations? University of Southern California, Information Sciences Institute.
25. Bunt, H., & Prasad, R. (2016). ISO DR-Core (ISO 24617-8): Core concepts for the annotation of discourse relations. In *Proceedings 12th joint ACL-ISO Workshop on Interoperable Semantic Annotation (ISA-12)* (pp. 45–54). Portoroz, Slovenia.
26. Chiarcos, C. (2014). Towards interoperable discourse annotation. Discourse features in the Ontologies of Linguistic Annotation. In *Proceedings of the Ninth International Conference on Language Resources and Evaluation (LREC'14)* (pp. 4569–4577).
27. Benamara, F., & Taboada, M. (2015). Mapping different rhetorical relation annotations: A proposal. In *Fourth Joint Conference on Lexical and Computational Semantics (* SEM 2015)* (pp. 147–152).
28. Roze, C., Braud, C., & Muller, P. (2019). Which aspects of discourse relations are hard to learn? Primitive decomposition for discourse relation classification. In *20th Annual SIGdial Meeting on Discourse and Dialogue (2019)* (pp. 432–441). ACL: Association for Computational Linguistics.

29. Fu, Y. (2023). Discourse relations classification and cross-framework discourse relation classification through the lens of cognitive dimensions: An empirical investigation. In *Proceedings of the 6th International Conference on Natural Language and Speech Processing (ICNLSP 2023)* (pp. 21–42).
30. Sanders, T. J., Spooren, W. P., & Noordman, L. G. (1992). Toward a taxonomy of coherence relations. *Discourse Processes, 15*(1), 1–35.
31. Sanders, T., Spooren, W., & Noordman, L. (1993). Coherence relations in a cognitive theory of discourse representation. *Cognitive Linguistics, 4*(2), 93–134. https://doi.org/10.1515/cogl.1993.4.2.93
32. Maat, H., & Sanders, T. (2000). Domains of use or subjectivity? The distribution of three Dutch causal connectives explained. In E. Couper-Kuhlen & B. Kortmann (Ed.), *Cause—Condition - Concession—Contrast: Cognitive and Discourse Perspectives* (pp. 57–82). De Gruyter Mouton. https://doi.org/10.1515/9783110219043.1.57
33. Rehbein, I., Scholman, M., & Demberg, V. (2016). Annotating discourse relations in spoken language: A comparison of the PDTB and CCR frameworks. In *Proceedings of the Tenth International Conference on Language Resources and Evaluation (LREC'16)* (pp. 1039–1046).
34. Pitler, E., Raghupathy, M., Mehta, H., Nenkova, A., Lee, A., & Joshi, A. (2008). Easily identifiable discourse relations. In *COLING 2008: Companion volume: Posters* (pp. 87–90).
35. Vaswani, A., Shazeer, N.M., Parmar, N., Uszkoreit, J., Jones, L., Gomez, A.N., Kaiser, L., & Polosukhin, I. (2017). Attention is all you need. In *Neural information processing systems*.
36. Gerych, W., Hartvigsen, T., Buquicchio, L., Agu, E., & Rundensteiner, E. A. (2021). Recurrent Bayesian classifier chains for exact multi-label classification. *Advances in Neural Information Processing Systems, 34*, 15981–15992.
37. Braud, C., Plank, B., & Søgaard, A. (2016). Multi-view and multi-task training of RST discourse parsers. In *Conference on Computational Linguistics (COLING)* (pp. 1903–1913).
38. Kim, N., Feng, S., Gunasekara, C., & Lastras, L. (2020). Implicit discourse relation classification: We need to talk about evaluation. In *Proceedings of the 58th Annual Meeting of the Association for Computational Linguistics* (pp. 5404–5414).
39. Reese, B., Hunter, J., Asher, N., Denis, P., & Baldridge, J. (2007). Reference manual for the analysis and annotation of rhetorical structure (version 1.0). *Technical report. University of Texas, Departments of Linguistics and Philosophy*. http://timeml.org/jamesp/annotation_manual.pdf
40. Wolf, T., Debut, L., Sanh, V., Chaumond, J., Delangue, C., Moi, A., ... & Rush, A. M. (2020). Transformers: State-of-the-art natural language processing. In *Proceedings of the 2020 Conference on Empirical Methods in Natural Language Processing: System Demonstrations* (pp. 38–45).
41. Paszke, A., Gross, S., Massa, F., Lerer, A., Bradbury, J., Chanan, G., ... & Chintala, S. (2019). Pytorch: An imperative style, high-performance deep learning library. In *Advances in neural information processing systems* (Vol. 32).
42. Braud, C., Liu, Y. J., Metheniti, E., Muller, P., Rivière, L., Rutherford, A. T., & Zeldes, A. (2023). The DISRPT 2023 shared task on elementary discourse unit segmentation, connective detection, and relation classification. In *3rd Shared Task on Discourse Relation Parsing and Treebanking (DISRPT 2023)* (pp. 1–21). ACL: Association for Computational Linguistics.
43. Cramér, H. (1999). *Mathematical methods of statistics* (Vol. 26). Princeton University Press.

POS Tagging as a Catalyst for Effective Named Entity Recognition in Low-Resource Languages

Mohamed Amine Cheragui, Abdelhalim Hafedh Dahou, Amin Abdedaiem, and Mourad Abbas

1 Introduction

NER is a cornerstone task in NLP, focusing on identifying and classifying entities such as names, locations, organizations, and dates within text [1]. This task plays a crucial role in many applications, including information retrieval, machine translation [2], sentiment analysis [3], document summarization [4], language understanding [5], and question-answering system [6]. However, achieving accurate NER performance becomes particularly challenging in under-resourced languages and dialects, due to linguistic variability, lack of standardized orthography, and limited annotated data. To address these challenges, researchers have explored leveraging additional linguistic features to enhance NER systems, one of which is POS tagging.

POS tagging, which involves assigning grammatical categories (e.g., nouns, verbs, adjectives) to each word in a sentence [7], is a foundational step in many NLP pipelines [8]. By providing syntactic context, POS tagging helps models better understand the relationships between words in a sentence, offering valuable information

M. A. Cheragui (✉) · A. Abdedaiem
Mathematics and Computer Science Department, Ahmed Draia University, Adrar, Algeria
e-mail: m_cheragui@univ-adrar.edu.dz; aminabdedaiem@gmail.com

A. H. Dahou
GESIS – Leibniz-Institute for the Social Sciences, Köln, Germany
e-mail: abdelhalim.dahou@gesis.org

M. Abbas
University of Science and Technology Houari Boumediene, Algiers, Algeria
e-mail: m_abbas04@yahoo.fr

M. Abbas (ed.), *Mastering Text Classification*, Signals and Communication Technology, https://doi.org/10.1007/978-3-031-93612-8_3

about the grammatical functions of words. This added context can complement NER by indicating which words are more likely to be named entities. For instance, proper nouns often appear as named entities, and POS tagging can help identify them with greater precision. While this integration has been explored in several languages, its application to dialectal Arabic, especially the Alg-D, remains underexamined.

This chapter investigates if integrating POS tagging can improve the performance of NER systems for the Alg-D when used as an extra feature. Specifically, we aim to verify the hypothesis that the inclusion of POS tagging enhances the ability of NER models to identify and classify named entities in a linguistically complex and low-resource context.

This study offers the following contributions to the field of NER in low-resource language processing:

- Demonstrates the impact of integrating POS tagging into NER models for under-resourced dialects.
- Provides a comprehensive evaluation of POS tags integration across diverse datasets, including Alg-D corpora (DzNER and NERDz) and modern standard Arabic (CLEANANERCorp).
- Proposes a novel pipeline that incorporates POS tags in dual representations (Arabic and Latin scripts), evaluating their effectiveness in dialectal NER tasks using this format word_pos-tag = ner-tag.
- Identifies limitations and provides a roadmap for future improvements in low-resource languages NER using linguistic features.

The remaining of this chapter is structured as follows: Sect. 2 reviews the background and related work, focusing on advancements in NER and the role of POS tagging in enhancing performance. Section 3 highlights the linguistic and resource-based challenges of applying NER to Alg-D. Section 4 outlines the experimental setup, including datasets, models, and evaluation metrics. Section 5 presents the results and discussion, demonstrating the impact of POS tagging on NER performance. Finally, Sect. 6 concludes the paper with a summary of findings and directions for future work.

2 Background and Related Work

The task of NER has long been a cornerstone in NLP, vital for extracting meaningful entities from text. Over the years, researchers have explored various methods to enhance NER performance, moving beyond traditional sequence labeling techniques. A compelling trend has emerged: leveraging linguistic features like POS tagging to improve both boundary detection and classification of entities. POS tagging offers syntactic context that complements semantic information, particularly valuable in morphologically rich or low-resource languages. This section delves into recent advancements that demonstrate the transformative impact of POS tagging on NER systems.

Cai et al. [9] recognized the challenges of entity boundary detection in Chinese and introduced a model that integrates POS tagging to tackle this issue. Their SM-LSTM-CRF model combines a self-matching attention layer, an LSTM layer, and a CRF layer, working together to enhance Chinese NER. To address errors from misclassified long entities, they proposed a reduced POS tagging method that focuses on general words, excluding potential medical terms. This approach yielded remarkable results, improving the F1 score by 2.59% with POS features and an additional 2.42% with the reduced tagging method, culminating in an overall F1 score of 80.07%.

In another effort, Suzuki et al. [10] simplified the Bi-LSTM-CRF architecture and harnessed transfer learning by fine-tuning the model on pretrained POS tagging tasks. Applied to Japanese datasets, their approach enhanced NER performance, achieving an F measure of 80.78% compared to 80.38% without fine-tuning, showcasing the benefits of syntactic information transfer.

Dou et al. [11] took a different path by introducing a unified model that integrates multi-grained word segmentation (MWS) and POS tagging within a two-stage joint parsing framework. This model leverages a hierarchical tree structure to share information across tasks, mitigating error propagation from word segmentation. Evaluated on OntoNotes datasets, it delivered substantial gains, with F1 scores of 82.82% on OntoNotes4 and 79.87% on OntoNotes5, compared to scores without POS tags.

For the specialized domain of Chinese Medical NER, Jiang et al. [12] developed the BBCPR model. This system employs a BERT-BiLSTM-CRF architecture augmented with a POS fusion layer and a novel regularization strategy named READ. By incorporating external syntactic knowledge, BBCPR achieved state-of-the-art F1 scores of 85.59% and 84.11% on the cMedQANER and cEHRNER datasets, respectively, significantly outperforming prior models.

Meanwhile, Güngór et al. [13] tackled NER in Turkish, a morphologically rich language, by jointly learning NER and morphological disambiguation (MD). Their Bi-LSTM-CRF-based joint model, J MULTI, successfully integrated morphological analyses, achieving an impressive F1 score of 83.21% for NER and demonstrating the efficacy of combined linguistic tasks.

For Telugu, Gorla et al. [14] utilized decision tree learning to generate POS rules, enhancing NER for a manually annotated corpus. Their system achieved F1 scores of 71.71% for person entities and 68.49% for locations, highlighting the utility of POS features in low-resource languages.

Arabic NER also benefited from morphological insights, as demonstrated by Farber et al. [15]. By integrating features from the morphological analysis and disambiguation for Arabic (MADA) system, they addressed challenges like lexical ambiguity and lack of capitalization, improving F1 scores from 66.7% to 71.5% on the ACE 2005 corpus.

Lastly, El-Karef et al. [16] proposed StagedNER, a two-stage fine-tuning framework that leverages POS tags to enhance Arabic NER. Their ensemble approach, using MSA POS tags, achieved an F1 score of 91.95%, outperforming the baseline by a wide margin and demonstrating the transformative potential of syntactic features in entity recognition.

These studies collectively underscore the critical role of POS tagging in advancing NER, especially in challenging linguistic contexts. By integrating syntactic cues with semantic models, researchers continue to push the boundaries of what NER systems can achieve.

3 NER Challenges for Algerian Dialects

Alg-D is a dynamic linguistic system shaped by centuries of interaction among Berber, French, Turc, Spanish, and Italian. While rooted in classical Arabic, Alg-D differs from it in terms of phonology, morphology, and syntax. This linguistic complexity, coupled with its informal nature, poses significant challenges for computational processing, particularly in tasks such as NER.

3.1 Lack of Standardization

Alg-D lacks standardization and normalization [17], which significantly impacts NER tasks, as the dialect exhibits high variability in spelling, vocabulary, and grammar due to influences from different languages. The absence of consistent linguistic rules leads to ambiguity in entity boundaries, inconsistent transliterations, and variations in how entities like names, places, and organizations are expressed across regions (Table 1).

Table 1 Examples of typographic variations of named entities in Alg-D

Word in MSA	NER label	1st variant	2nd variant	3rd variant	Translation
الجزائر	LOC	العاصمة	دزاير	الحي	Algiers
سوناطراك	ORG	سونطراك	سوناتراك	صوناطراك	SONATRACH
محامي	PER	بوقاطو	بوخاطو	أفوكا	Lawyer

Furthermore, the informal and primarily oral nature of the dialect makes it difficult to create annotated datasets with unified entity representations, hindering the training and performance of NER models. However, the presence of CODA (conventional orthography for dialectal Arabic) offers a promising solution by providing a standardized writing system for dialectal Arabic [18].

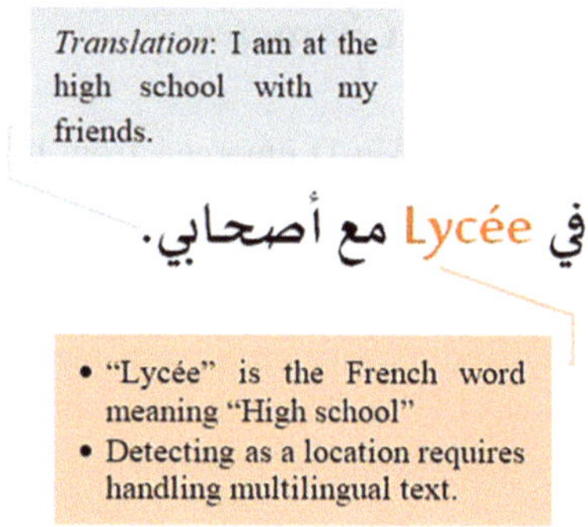

Fig. 1 Code-switching phenomena in Alg-D

3.2 Code Switching

Code switching is the linguistic phenomenon where speakers alternate between two or more languages or dialects within a single conversation, sentence, or even a phrase [19]. This behavior is common in multilingual societies like Algeria, where historical, social, and cultural influences have led to the coexistence of Arabic, French, Berber, and, more recently, English. Identifying entities in such multilingual contexts is challenging [20] (Fig. 1).

3.3 Agglutination

MSA is inherently an agglutinative language, where words are formed by attaching affixes to a root, each affix contributing specific grammatical or semantic information [21]. The Alg-D, as a variant of MSA, has inherited this characteristic but takes it further by fusing pronouns, prepositions, and other elements into single compact forms. While this linguistic trait streamlines spoken communication, it poses significant challenges for tasks such as NER. The agglutination in Alg-D often obscures clear boundaries between entities and their context, making it more difficult to isolate and classify names, places, or other entities accurately, especially when compared to the more structured and explicit morphology of MSA (Table 2).

Table 2 Illustrative cases of agglutination in Alg-D

Sentance	Entity	NER label	Affix	Translation
تلاقيت بمحمد البارح	بمحمد	PER	بـ	I met Mohamed yesterday
يسكن قريب لسيفيتال	لسيفيتال	ORG	لـ	Living close to Cevital
وصلتها لسبيطارها	لسبيطارها	LOC	لـ / ها	I accompanied her to the hospital

3.4 Integration of Non-Arabic Characters

The adoption of the Latin script in Alg-D emerged from historical and practical contexts [22], such as the widespread use of Latin-script-based devices and platforms. This shift gave rise to Arabizi, a transliteration system where Alg-D is written phonetically using Latin letters and numerals, and Arabizi is widely used in informal communication, particularly on social media and messaging apps. However, its inconsistent orthography and informal nature pose challenges for NER tasks, which aim to identify and classify proper nouns like names of people, organizations, and locations. The application of NER to Arabizi-transliterated texts requires robust models capable of handling noisy data and linguistic variability, making it an intriguing and complex problem in NLP (Table 3).

Table 3 Examples of the Arabizi phenomenon in Alg-D

Word in MSA	NER label	1st variant	2nd variant	3rd variant	Translation
خبير	PER	Khabir	5bir	5abir	Expert
وهران	LOC	Wahrn	Wahrane	Ouahran	Oran
بريد الجزائر	ORG	Labosta	El-Bosta	La poste	Algerian Post

3.5 Limited Resources in Alg-D

The rarity of annotated datasets for Alg-D has been a significant challenge to advance NLP tasks [23], particularly NER. However, the introduction of resources like DzNER [24] and NERDz [25] marks a significant step forward in addressing this gap. These datasets provide valuable annotated examples that help capture the unique linguistic characteristics of Alg-D and serve as foundational resources for developing and evaluating NER models.

3.6 Ambiguity of Named Entities

Ambiguity in named entities arises primarily from linguistic challenges such as homographs, where words share identical spelling but differ in meaning, and the absence of diacritics in written text, which obscures pronunciation and semantic interpretation, particularly in MSA [26]. Many personal names are either derived from adjectives or can be mistaken for other nouns with the same surface form. Additionally, some Arabic personal names coincide with verbs or prepositions [27]. This issue is further compounded in the Alg-D, which introduces additional complexity with composed entities that often carry multiple interpretations, depending on the context. Moreover, cultural and historical references in Alg-D often blur the boundaries between proper nouns and colloquial expressions, making it difficult for NER systems to accurately identify and classify entities without robust contextual understanding (Table 4).

Table 4 Examples of named entity ambiguities in Alg-D

Ambiguous example	Incorrect label	Correct label	Translation
نحب نشرب حمود بوعلام	PER	ORG	I like to drink Hamoud Boualem
لالة فاطمة نسومر	PER	Out	Lalla Fatma N'Soumer
صاحبي يسكن فسيدي جيلالي	PER	LOC	My friend lives in Sidi Djilali

4 Experimental Methodology

This section presents our experimental methodology for evaluating the impact of POS tagging on Arabic NER. Our approach augments token classification by incorporating POS features into an Arabic pretrained language model to assess potential performance enhancements in NER tasks.

4.1 Datasets

Our experimental evaluation utilized two Alg-D datasets alongside one modern standard Arabic dataset:

- **DzNER**, a corpus constructed for Alg-D NER, contains 220,000 tokens covering Sports, Travel, Electronics, and Political domains. The corpus features 18,387 annotated entities across three categories: organization (ORG), person (PER), and location (LOC). The dataset follows an 80-20 split for training and testing, respectively.

- **NERDz**, an Alg-D NER corpus based on NArabizi treebank [28], contains 19,258 tokens with parallel text in Latin and Arabic writing styles. The corpus encompasses eight entity categories: PER (467), GPE[1] (438), ORG (290), NORP[2] (235), EVT[3] (54), LOC (41), PROD[4] (23), and MISC[5] entities (18). In our study, we used the version written in Arabic writing style and took just the train and test splits to keep our study comparable with [25].
- **CLEANANERCorp** [29], for MSA NER, we utilized the cleaned version of the original ANERcorp dataset [30]. CLEANANERcorp consists of 316 articles selected from different newspapers to create a diverse corpus, totaling 150,110 tokens. The training split comprises 125,102 tokens, and the test split contains 25,008 tokens, all labeled with ORG, PER, LOC, and MISC tags (Table 5).

Table 5 Entity counts in different datasets

Type	NERDz	DzNER	CLEANANERCorp
PER	408	6159	6410
GPE	383	–	–
ORG	268	2312	5260
NORP	206	–	–
EVT	49	–	–
LOC	38	5164	5462
PROD	16	–	–
MISC	18	–	6359

4.2 Baselines

In our study, due to the lack of POS-tag-enhanced models for MSA and dialectal Arabic, we used Arabic pretrained models based on Huggingface Transformers library and examined the impact of injecting POS tags into the input. Model selection was based on specifications like vocabulary size, parameter count, language diversity in training data, and the ability to generalize across related NLP classification tasks.

[1] GPE: Primarily signifies countries but also includes entities with parliamentary-like governance systems, encompassing states and cities as well, which are categorized as GPEs.

[2] NORP: Refers to groups of individuals who share a common nationality, political ideology, religion, or are identified by proper nouns representing fans of football clubs.

[3] EVT: This category primarily pertains to sports events and political elections.

[4] PROD: Represents objects or a series of objects, provided they are human-made, such as TVs and vehicles.

[5] MISC: Includes all entities that are infrequent in our dataset, such as quantities, money, diseases, and chemical components.

Based on these criteria, we evaluated multiple Arabic pretrained models as described in Fig. 2. First, we considered models designed for MSA, including ARBERT [31] and CAMeLBERT_MSA [32]. Next, we examined models specifically trained on Arabic dialects, such as MARBERTv2 [31], DziriBERT [33], and CAMeLBERT_DA [32]. Finally, we reviewed models that encompass both MSA and dialects, including AraBERTv2 [34], Qarib [35], XLM-RoBERTa-base [36], and multilingual-cased BERT (mBERT-cased) [37] to determine the most suitable model for our task, one that can handle nuanced linguistic understanding.

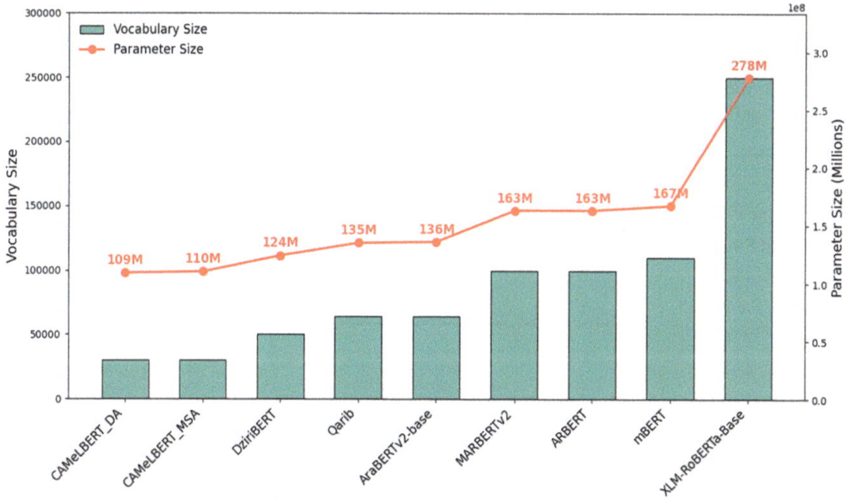

Fig. 2 Comparison between the used Arabic pretrained models in terms of vocabulary and parameter size

4.3 Procedure

In this section, we will describe the different stages we went through that helped us investigating the impact of POS on Arabic NER. The pipeline illustrated in Fig. 3 is composed of the following processes: POS tag extraction, data preparation, and model fine-tuning:

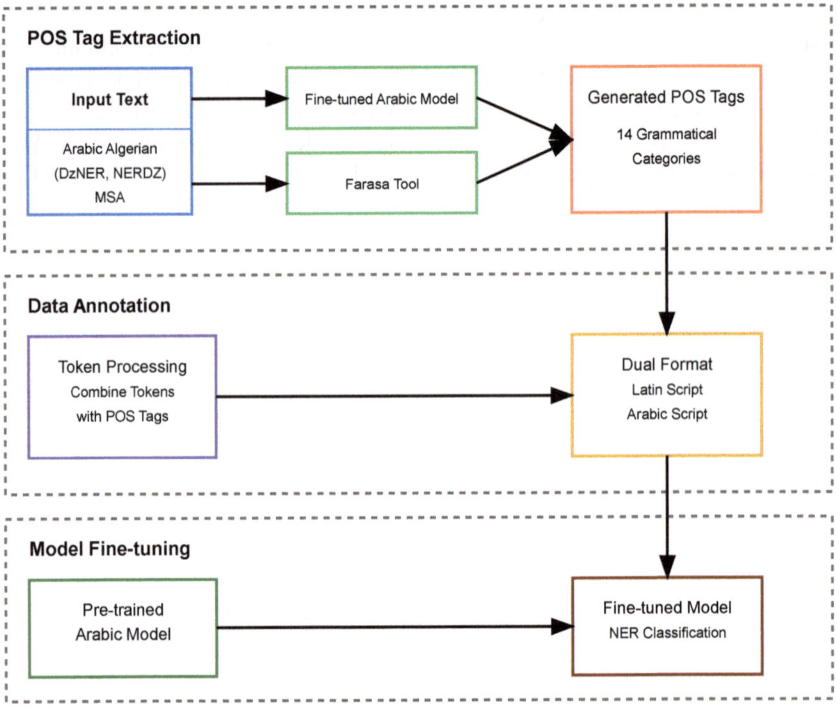

Fig. 3 A comprehensive pipeline illustrating the three-stage process on investigating the impact of POS on Arabic NER passing by POS tag extraction, data preparation, and model fine-tuning

- **POS Tag Extraction** For each token in a given Arabic Alg-D sentence (DzNER and NERDz), a grammatical category was extracted using a pretrained and fine-tuned Arabic language model [38]. The POS tag extractor was specifically trained on the NArabizi dataset described in [28]. This dataset comprises 1300 Arabizi sentences sourced from an Algerian newspaper's web forum and an additional 200 sentences manually collected from song lyrics on various online platforms. The dataset was labeled using 14 POS tags. For datasets written in MSA (CLEANANERCorp), we employed Farasa [39], a state-of-the-art POS tagging tool for MSA with 83% accuracy. Farasa provides POS tagging coverage for the same grammatical categories as shown in Table 6, ensuring consistency across both Alg-D and MSA datasets. These tools analyze input sentences and generate POS tags by considering the full context, guaranteeing accurate and contextually informed grammatical classifications.

Table 6 POS Tags with their Arabic correspondences and English translations

POS Tag	Arabic	English
NOUN	اسم	Noun
VERB	فعل	Verb
ADJ	صفة	Adjective
PROPN	اسم عَلَم	Proper Noun
PRON	ضمير	Pronoun
ADV	حال او ظرف	Adverb
ADP	حرف جر	Adposition
INTJ	أداة تعجب او انفعال	Interjection
DET	أداة تعريف	Determiner
SCONJ	أداة ربط ثانوية	Subordinating Conjunction
PART	أداة	Particle
PUNCT	علامة ترقيم	Punctuation
CCONJ	حرف عطف	Coordinating Conjunction
NUM	عدد	Numeral
X	مجهول	Unknown

- **Data preparation** The extracted POS tags were concatenated with their corresponding tokens and NER labels. Specifically, the NER labels for each token were sourced directly from the dataset annotations. The output format for each token was structured as a word paired with its associated NER label (Word_POS). This step ensured the incorporation of POS information alongside the original token for downstream tasks. To explore the potential impact of script representation, we generated two versions of the dataset: one where POS tags were retained in their original Latin script (as used in the NArabizi dataset) and another where the POS tags were transliterated into Arabic script. This dual representation was designed to assess the performance of models in these two scenarios, recognizing that most pretrained models are optimized for Arabic script rather than Latin script.
- **Model Fine-Tuning** The enriched dataset, containing tokens with POS information, was used to fine-tune a selection of pretrained Arabic language models. These models were trained to classify each token in a sentence into one of the predefined NER categories. The inclusion of POS tags as supplementary information aimed to improve the model's understanding of the syntactic and semantic properties of the tokens, thus enhancing the NER classification performance.

4.4 Hyperparameter Configuration

Our evaluation experiments were conducted on the Google Colab platform utilizing an A100-40GB GPU. For the DzNER and CLEANANERCorp datasets, we employed consistent hyperparameter settings across all models: the Adam optimizer [40] with a learning rate of 5×10^{-5}, a batch size of 16, a seed value of 42, six training epochs, and warmup steps of 21. For the NERDz dataset, we maintained the hyperparameters specified in the original paper, including a learning rate of 5×10^{-5}, five training epochs, and a batch size of 8. All experiments were implemented using the Huggingface Transformers library [41] (Table 7).

Table 7 Hyperparameters configuration for models fine-tuning

Hyperparameter	DzNER	CLEANANERCorp	NERDz
Learning rate	5×10^{-5}	5×10^{-5}	5×10^{-5}
Batch size	16	16	8
Number of epochs	6	6	5
Optimizer	Adam	Adam	Adam
Warmup steps	21	21	100
Random seed	42	42	42
Weight decay	0.03	0.03	0.01

5 Results and Discussion

5.1 NERDz Dataset

Table 8 reveals the performance of the Arabic language models evaluated in the NERDz dataset with and without POS tag integration scenarios. AraBERTv02 demonstrated optimal balanced performance, achieving the highest precision (76.19%) and recall (81.52%) with POS tags while maintaining a consistent F1 score of 78.76%. MARBERTv2 showed marked improvement with POS tags, reaching 97.07% accuracy and significantly increased recall, though with some precision trade-offs. Interestingly, QARIB performed better without POS tags, suggesting its pretraining effectively captures syntactic information independently. The XLM-RoBERTa-base showed comprehensive improvements with POS integration, suggesting that large (in terms of parameters and vocabulary size) and multilingual models can benefit significantly from explicit syntactic information. DziriBERT and CamelBERT_da displayed minimal sensitivity to POS information, which might indicate that these models' architectures or pretraining approaches and sizes make them less sensitive to explicit POS information. Overall, while POS tag integration generally enhanced model performance, particularly in recall, the benefits varied substantially across different model architectures.

Table 8 Performance metrics comparison for models on NERDz dataset with/without POS tags

Model	Accuracy		Precision		Recall		F1	
	POS	No POS	POS	No POS	POS	No POS	POS	No POS
XLM-RoBERTa-base	96.40	95.37	73.45	68.38	75.79	67.51	74.60	67.94
QARIB	96.40	96.83	78	83.09	74.52	75.15	76.22	78.92
CamelBERT_da	95.66	95.65	66.86	70.96	73.24	70.06	69.90	70.51
MARBERTv2	97.07	96.41	74.09	76.35	78.34	71.97	76.16	74.09
AraBERTv02	96.98	96.83	76.19	85.18	81.52	73.24	78.76	78.76
mBERT	96.62	96.17	75.92	73.88	78.34	73.88	77.11	73.88
DziriBERT	95.90	96.50	68.45	75.49	73.24	72.61	70.76	74.02

The entity-level analysis as shown in Fig. 4 reveals notable disparities in recognition performance across different entity types. GPE and PER consistently demonstrated robust recognition capabilities, achieving F1 scores above 80% across both configurations. This superior performance can be attributed to their well-defined contextual patterns and distinctive features in Arabic text. However, identifying LOC entities was especially difficult when utilizing POS tags, which resulted in a failure in identification (F1 = 0%), while without using them had a limited success rate (F1 = 28.57%). This significant contrast raises the possibility that noise introduced by existing POS integration that may hinder location entity identification, especially with a low number of samples in the train set. For LOC elements with numerical components, like (20 أوت or "20 August"), this problem is more noticeable because the POS tag differs from "proper noun" or "noun, " leading the system to mistakenly classify them as non-location entities. Furthermore, the problem is also present in sentences that involve a lot of code-switching words, as the combination of languages (for example, Arabic and French or English) makes it more difficult to accurately assign correct POS tags, which can result in incorrect categorization or a failure to recognize LOC entities.

PROD mentions showed marked improvement with POS integration, with F1 scores increasing from 50% to 67%, indicating that syntactic features play a crucial role in distinguishing product references from general nouns. Conversely, ORG recognition showed moderate degradation with POS integration, while EVT recognition remained notably consistent regardless of POS information, maintaining an F1 score of 73%. These findings suggest that the utility of POS tags in Arabic NER is highly entity dependent, especially in terms of quantity inside dataset, warranting careful consideration in model design and feature selection.

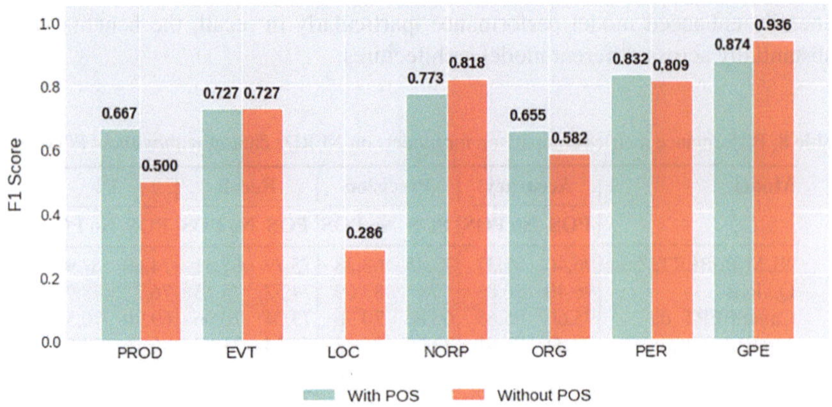

Fig. 4 Comparison of best models on each entity in the NERDz, with vs. without POS features. Notably, the LOC entity's score drops to zero with POS

5.2 DzNER Dataset

The analysis of performance metrics across various models on the DzNER dataset is presented in Table 9. The integration of POS tags resulted in significant accuracy gains across all models. AraBERTv02 exhibited the highest improvement, with an accuracy increase from 95.70% to 97.38%. This suggests that the syntactic information provided by POS tagging is particularly effective in enhancing the overall classification correctness of the model. Precision, which measures the correctness of identified entities, showed improvements in almost all models with the addition of POS tags. MARBERTv2 showed the most substantial precision gain, increasing from 86.71% to 91.09%. This indicates that POS tags help the model more accurately delineate entity boundaries and distinguish between entity and nonentity tokens. Regarding recall, which assesses the model's ability to identify all relevant entities, there was a notable improvement in AraBERTv02, jumping from 66.55% to 81.55%. This improvement underscores the benefits of POS tagging in helping the model capture more entities without increasing false positives. F1 score was consistently higher when POS tags were used over all models which highlight that POS tagging not only increases the accuracy of entity recognition but also makes the models more reliable by improving both precision and recall.

While all models benefited from the inclusion of POS tags, the extent of improvement varied. Models like AraBERTv02 and MARBERTv2 showed the most significant gains, which may be attributed to their underlying size and training methodologies that make more effective use of syntactic information. In contrast, models like DziriBERT, despite showing overall improvements, had a mixed response in terms of precision. This variability underscores the importance of model-specific tuning and adaptation when incorporating linguistic features like POS tagging.

Table 9 Performance metrics comparison for models on DzNER dataset with/without POS tags

Model	**Accuracy**		**Precision**		**Recall**		**F1**	
	POS	**No POS**	**POS**	**No POS**	**POS**	**No POS**	**POS**	**No POS**
XLM-RoBERTa-base	96.85	94.47	84.62	83.67	79.17	60.95	81.80	70.53
QARIB	96.66	94.76	85.12	83.55	75.62	60.83	80.09	70.40
CamelBERT_da	96.32	94.48	83.22	74.09	73.33	59.86	77.96	66.22
MARBERTv2	97.11	95.81	91.09	86.71	77.85	65.26	83.95	74.47
AraBERTv02	97.38	95.70	87.97	83.25	81.55	66.55	84.64	73.97
mBERT	96.58	95.09	84.35	78.31	75.45	62.54	79.65	69.54
DziriBERT	96.91	94.62	86.17	86.88	77.90	54.79	81.83	67.20

Figure 5 shows the F1 score comparisons on the DzNER dataset across all categories. The performance with POS tagging (F1 score of 76.9%) on the ORG category significantly outperforms the model without POS tagging (F1 score of 64.5%). This indicates that POS tagging is particularly beneficial in correctly identifying and classifying organizational names, which might be due to better disambiguation of nouns and proper nouns when POS tags are used. For LOC, we observe that the impact of POS tagging is even more pronounced, with the F1 score increasing from 81.7% to 89.4% when POS tags are incorporated. This suggests that POS tagging aids immensely in recognizing geographical entities, likely by providing contextual cues that differentiate location names from other entities. Lastly, for the main dominant category PER, the use of POS tags results in an F1 score of 85.9% compared to 74.9% without POS tags. While the improvement is noticeable, it is less dramatic than in the LOC category, especially in terms of distribution inside the dataset, possibly because personal names have fewer contextual linguistic patterns that POS tagging can leverage compared to locations.

5.3 CLEANANERCorp Dataset

The impact of POS tagging on MSA NER was investigated using pretrained models and reported in Table 10. The results showed that incorporating POS tags did not consistently improve performance across all metrics compared to the previous two datasets. While models with POS tags achieved high accuracy, they generally exhibited lower precision, recall, and F1 scores compared to models trained without POS tags. The best-performing model, AraBERTv02, achieved an F1 score of 89.96% without POS tags, which was paradoxically slightly higher than its F1 score of 88.64% with POS tags. This suggests that POS information might not be as beneficial as expected for this specific language and could potentially introduce noise or redundancy into the learning process.

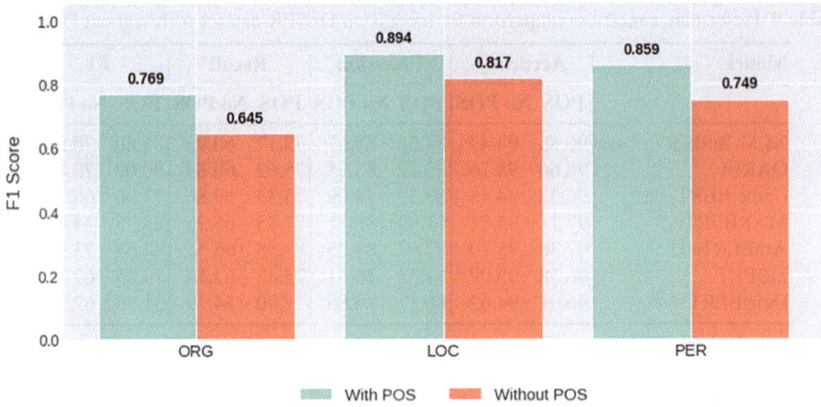

Fig. 5 Comparison of best models on each entity in the DzNER, with vs. without POS features

Table 10 Performance metrics comparison for models on ANERCorp dataset with/without POS tags

Model	**Accuracy**		**Precision**		**Recall**		**F1**	
	POS	**No POS**	**POS**	**No POS**	**POS**	**No POS**	**POS**	**No POS**
XLM-RoBERTa-base	96.44	96.81	83.27	85.41	82.59	84.10	82.93	84.75
QARIB	95.42	96.85	77.51	84.06	79.64	85.63	78.56	84.84
CamelBERT_MSA	97.25	97.76	87.67	89.01	87.73	89.75	87.70	89.38
AraBERTv02	97.46	97.80	88.53	89.76	88.75	90.15	88.64	89.96
mBERT	95.68	95.97	80.60	82.73	79.87	79.90	80.23	81.29
ARBERTv02	97.33	97.74	88.55	89.47	87.65	89.62	88.10	89.54

A closer examination of the F1 scores for different entity types as shown in Fig. 6 revealed varying degrees of impact. For ORG and LOC, the F1 scores were comparable for models with and without POS tags, indicating that POS information might not be crucial for these entities. However, for MISC entities, the model trained without POS tags achieved an F1 score of 89.38%, which was significantly higher than the F1 score of 87.70% with POS tags. Similarly, for PER entities, the model without POS tags achieved an F1 score of 92.1%, outperforming the model with POS tags which achieved an F1 score of 91.5%. This suggests that POS information might be particularly detrimental for these entity types, potentially due to the introduction of noise or conflicting information with what seen during pretraining phase and fine-tuning phase.

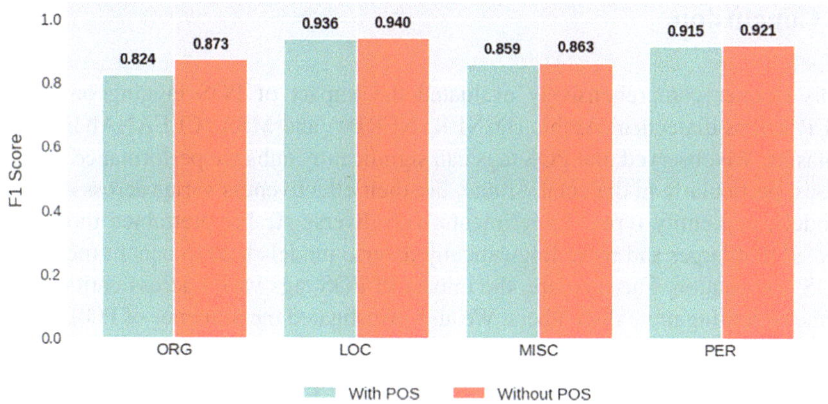

Fig. 6 Comparison of best models on each entity in the CLEANANERCorp, with vs. without POS features

Furthermore, using POS tags in Latin may have added unnecessary complexity, potentially limiting the model's ability to fully leverage POS information for improved understanding. To investigate whether the POS format directly impacts the NER task, an ablation study was conducted using Arabic POS tags, allowing for a more intuitive integration of linguistic features. The average F1 score of the best model with Arabic POS tags is 88.7%, while the average F1 score of best model with Latin POS tags is 88.64% as shown in Table 11. This indicates that using Arabic POS tags results in slightly better performance (as observed with models like AraBERTv02, CamelBERT_msa, and ARBERTv02, which are trained on only Arabic language) or at least does not negatively impact the performance of pretrained models on the CLEANANERCorp dataset. The results underscore the potential of integrating linguistically relevant information, such as Arabic POS tags, into NER models to enhance their performance, particularly when the models are trained on and applied to the same language.

Table 11 CLEANANERCorp with Arabic POS tags performance metrics

Model	Accuracy	Precision	Recall	F1 score
XLM-RoBERTa-base	96.26	83.41	82.10	82.75
QARIB	95.42	76.67	79.74	78.17
CamelBERT_msa	97.38	88.53	88.65	88.59
ARBERTv02	97.32	88.11	88.44	88.27
mBERT	95.90	81.04	79.98	80.51
AraBERTv02	97.46	88.46	88.94	88.70

6 Conclusion

This chapter comprehensively evaluated the impact of POS tagging on Arabic NER across dialectical Arabic (DzNER, NERDz) and MSA (CLEANANERCorp) datasets. We observed that POS tags can significantly enhance performance in some cases, particularly in dialectal Arabic, but their effectiveness varies across datasets, models, and entity types. Experiments with diverse Arabic pretrained models revealed that larger and more linguistically diverse models tend to benefit more from POS information. Furthermore, the impact of POS tags varies across entities, with some benefiting more than others. We also investigated the influence of POS tag representation, finding that using Arabic-script POS tags yields slightly better results compared to Latin-script tags, even though the improvement is negligible but further investigation as well as fine-tuning LLM's can either prove or deny its benefits. These findings underscore the importance of considering the specific characteristics of the dataset, model architecture, entity types, and POS tag representation when incorporating POS information into Arabic NER systems.

As a future work, we will concentrate on:

- Exploring alternative POS tag representations and strategies of injection
- Investigating the impact of POS tag quality by using different POS taggers
- Mixing multilingual datasets and test them on a multilanguage LLM's to further prove and remove any suspicions about language LLM's biases
- Analyzing the interaction between POS tags and other linguistic features, such as morphological information or contextual embeddings offered by large language models (LLMs)

References

1. Qu, X., Gu, Y., Xia, Q., Li, Z., Wang, Z., & Huai, B. (2023). A survey on Arabic named entity recognition: Past, recent advances, and future trends. *IEEE Transactions on Knowledge and Data Engineering, 36*(3), 943–959.
2. Ugawa, A., Tamura, A., Ninomiya, T., Takamura, H., & Okumura, M. (2018). Neural machine translation incorporating named entity. *Proceedings Of The 27th International Conference On Computational Linguistics* (pp. 3240–3250).
3. Mulki, H., Haddad, H., Gridach, M., & Babaoglu, I. (2019). Empirical evaluation of leveraging named entities for Arabic sentiment analysis. ArXiv Preprint ArXiv:1904.10195.
4. Khan, I., Sheikh, A., & Sinha, U. (2024). Graph neural network and NER-based text summarization. ArXiv Preprint ArXiv:2402.05126.
5. Pakhale, K. (2023). Comprehensive overview of named entity recognition: Models, domain-specific applications and challenges. ArXiv Preprint ArXiv:2309.14084.
6. Toral, A., Noguera, E., Llopis, F., & Munoz, R. (2005). Improving question answering using named entity recognition. In *Natural Language Processing and Information Systems: 10th International Conference on Applications of Natural Language to Information Systems, NLDB 2005, Alicante, Spain, June 15–17, 2005. Proceedings 10* (pp. 181–191).
7. Nerabie, A., AlKhatib, M., Mathew, S., El Barachi, M., & Oroumchian, F. (2021). The impact of Arabic part of speech tagging on sentiment analysis: A new corpus and deep learning approach. *Procedia Computer Science, 184*, 148–155.

8. Chiche, A., & Yitagesu, B. (2022). Part of speech tagging: A systematic review of deep learning and machine learning approaches. *Journal of Big Data, 9*, 10.
9. Cai, X., Dong, S., & Hu, J. (2019). A deep learning model incorporating part of speech and self-matching attention for named entity recognition of Chinese electronic medical records. *BMC Medical Informatics And Decision Making, 19*, 109.
10. Suzuki, M., Komiya, K., Sasaki, M., & Shinnou, H. (2018). Fine-tuning for named entity recognition using part-of-speech tagging. In *Proceedings Of The 32nd Pacific Asia Conference On Language, Information And Computation.*
11. Dou, C., Gong, C., Li, Z., Wang, Z., Huai, B., & Zhang, M. (2024). Improving Chinese named entity recognition with multi-grained words and part-of-speech tags via joint modeling. In *Proceedings of the 2024 Joint International Conference on Computational Linguistics, Language Resources and Evaluation (LREC-COLING 2024)* (pp. 8732–8742).
12. Jiang, M., Zhang, X., Chen, C., Shao, T., & Chen, H. (2022). Leveraging part-of-speech tagging features and a novel regularization strategy for Chinese medical named entity recognition. *Mathematics, 10*, 1386.
13. Güngör, O., Uskudarli, S., & Güngör, T. (2018). Improving named entity recognition by jointly learning to disambiguate morphological tags. In *Proceedings of the 27th International Conference on Computational Linguistics* (pp. 2082–2092). https://aclanthology.org/C18-1177
14. Gorla, S., Velivelli, S., Satpathi, D., Murthy, N., & Malapati, A. (2020). Named entity recognition using part-of-speech rules for Telugu. In *Smart computing paradigms: new progresses and challenges* (pp. 147–157).
15. Farber, B., Freitag, D., Habash, N., & Rambow, O. (2008). Improving NER in Arabic using a morphological tagger. In *LREC.*
16. Elkaref, N., & El-karef, M. (2023). El-Kawaref at WojoodNER shared task: StagedNER for Arabic named entity recognition. In *Proceedings of ArabicNLP 2023* (pp. 803–808).
17. Bousmaha, K., Hamadouche, K., Gourara, I., & Hadrich, L. (2022). DZ-OPINION: Algerian dialect opinion analysis model with deep learning techniques. *Revue D'Intelligence Artificielle, 36*, 897.
18. Saadane, H., & Habash, N. (2015). A conventional orthography for Algerian Arabic. In *The Second Workshop on Arabic Natural Language Processing* (pp. 69–79).
19. Winata, G., Aji, A., Yong, Z., & Solorio, T. (2022). The decades progress on code-switching research in NLP: A systematic survey on trends and challenges. ArXiv Preprint ArXiv:2212.09660.
20. Aguilar, G., AlGhamdi, F., Soto, V., Diab, M., Hirschberg, J., & Solorio, T. (2019). Named entity recognition on code-switched data: Overview of the CALCS 2018 shared task. ArXiv Preprint ArXiv:1906.04138.
21. Shaalan, K., & Raza, H. (2009). NERA: Named entity recognition for Arabic. *Journal of the American Society for Information Science and Technology, 60*, 1652–1663.
22. Dahou, B., Falek, L., Abbas, M., Mekaoui, S., Lichouri, M., & Zitouni, A. (2023). DZ-SMS: an authentic corpus of Algerian SMS. *ACM Transactions On Asian And Low-Resource Language Information Processing, 22*, 1–21.
23. Moudjari, L., Akli-Astouati, K., & Benamara, F. (2020). An Algerian corpus and an annotation platform for opinion and emotion analysis. In *Proceedings of the Twelfth Language Resources and Evaluation Conference* (pp. 1202–1210).
24. Dahou, A., & Cheragui, M. (2023). Dzner: A large Algerian named entity recognition dataset. *Natural Language Processing Journal, 3*, 100005.
25. Touileb, S. (2022). Nerdz: A preliminary dataset of named entities for Algerian. In *Proceedings of the 2nd Conference of the Asia-Pacific Chapter of the Association for Computational Linguistics and the 12th International Joint Conference on Natural Language Processing (Volume 2: Short Papers)* (pp. 95–101).
26. Benajiba, Y., Diab, M., & Rosso, P. (2008). Arabic named entity recognition using optimized feature sets. In *Proceedings of the 2008 Conference on Empirical Methods in Natural Language Processing* (pp. 284–293).

27. Zayed, O., & El-Beltagy, S. (2015). Named entity recognition of persons' names in Arabic tweets. In *Proceedings of the International Conference Recent Advances in Natural Language Processing* (pp. 731–738).
28. Touileb, S., & Barnes, J. (2021). The interplay between language similarity and script on a novel multi-layer Algerian dialect corpus. ArXiv Preprint ArXiv:2105.07400.
29. Al-Duwais, M., Al-Khalifa, H., & Al-Salman, A. (2024). CLEANANERCorp: Identifying and correcting incorrect labels in the ANERcorp dataset. ArXiv Preprint ArXiv:2408.12362.
30. Benajiba, Y., Rosso, P., & Benediruiz, J. (2007). Anersys: An arabic named entity recognition system based on maximum entropy. In *Computational Linguistics and Intelligent Text Processing: 8th International Conference, CICLing 2007, Mexico City, Mexico, February 18–24, 2007. Proceedings 8* (pp. 143–153).
31. Abdul-Mageed, M., Elmadany, A., & Nagoudi, E. (2020). ARBERT & MARBERT: Deep bidirectional transformers for Arabic. ArXiv Preprint ArXiv:2101.01785.
32. Inoue, G., Alhafni, B., Baimukan, N., Bouamor, H., & Habash, N. (2021). The interplay of variant, size, and task type in Arabic pre-trained language models. ArXiv Preprint ArXiv:2103.06678.
33. Abdaoui, A., Berrimi, M., Oussalah, M., & Moussaoui, A. (2021). Dziribert: A pre-trained language model for the algerian dialect. ArXiv Preprint ArXiv:2109.12346.
34. Antoun, W., Baly, F., & Hajj, H. (2020). Arabert: Transformer-based model for arabic language understanding. ArXiv Preprint ArXiv:2003.00104.
35. Abdelali, A., Hassan, S., Mubarak, H., Darwish, K., & Samih, Y. (2021). Pre-Training BERT on Arabic Tweets: Practical considerations.
36. Conneau, A. (2019). Unsupervised cross-lingual representation learning at scale. ArXiv Preprint ArXiv:1911.02116.
37. Devlin, J. (2018). Bert: Pre-training of deep bidirectional transformers for language understanding. ArXiv Preprint ArXiv:1810.04805.
38. Cheragui, M., Dahou, A., & Abdedaiem, A. (2023). Exploring BERT models for part-of-speech tagging in the algerian dialect: A comprehensive study. In *Proceedings of the 6th International Conference on Natural Language and Speech Processing (ICNLSP 2023)* (pp. 140–150).
39. Abdelali, A., Darwish, K., Durrani, N., & Mubarak, H. (2016). Farasa: A fast and furious segmenter for Arabic. In *Proceedings of the 2016 Conference of the North American Chapter of the Association for Computational Linguistics: Demonstrations* (pp. 11–16).
40. Kingma, D., & Ba, J. (2014). Adam: A method for stochastic optimization. ArXiv Preprint ArXiv:1412.6980.
41. Wolf, T., Debut, L., Sanh, V., Chaumond, J., Delangue, C., Moi, A., Cistac, P., Rault, T., Louf, R., Funtowicz, M., et al. (2020). Transformers: State-of-the-art natural language processing. In *Proceedings of the 2020 Conference on Empirical Methods in Natural Language Processing: System Demonstrations* (pp. 38–45).

QdGF: A Framework Which Leverages Training-Time Information for Enhanced Multilingual-Multitask Classification

Sebastian-Antonio Toma and Camelia Lemnaru

1 Introduction

Classification is among the most prevalent applications of artificial intelligence, finding extensive use across diverse domains. Text classification, in particular, is a cornerstone task in natural language processing, underpinning a range of applications such as sentiment analysis, topic labeling, question answering, and recipe classification, even enabling smarter solutions like optimizing instructions for modern kitchen appliances. In an age defined by an overwhelming abundance of textual data driven by the increasing diversity and complexity of information generated across various websites in various languages, manual processing and categorization are not only labor intensive but also inefficient [1]. This challenge underscores the growing necessity of leveraging deep learning techniques to automate and tackle more complex classification tasks effectively.

S.-A. Toma (✉) · C. Lemnaru
Computer Science Department, Technical University of Cluj-Napoca, Cluj-Napoca, Romania
e-mail: sebastianantoniotoma@gmail.com; Camelia.Lemnaru@cs.utcluj.ro

M. Abbas (ed.), *Mastering Text Classification*, Signals and Communication Technology, https://doi.org/10.1007/978-3-031-93612-8_4

Beyond its standalone applications, text classification can also be integrated within a multitask learning (MTL) framework. MTL aims to enhance the capabilities of natural language processing models by enabling them to learn multiple interrelated tasks simultaneously, leveraging shared representations to boost overall performance and efficiency. According to recent studies [2], MTL has proven particularly effective in reducing overfitting, improving generalization across tasks, and enabling the transfer of knowledge from high-resource to low-resource settings, making it a valuable approach in multilingual NLP scenarios.

This chapter continues our work in [3] on the application of MTL techniques in the context of multilingual text classification, with the declared goal of leveraging the inherent relationships between different classification tasks to improve the accuracy and robustness of the model used. We further extend the application of QdGF to other multilingual generative models as backbones, and we focus on the same three classification tasks, where the labels represent cooking properties for oven recipes. We first show the improvement brought by implementing joint learning (a special form of MTL, using all three main tasks), followed by introducing a more complex related task while learning, and then finally we integrate the questions-based generative framework. For this framework, an ablation study will be resurfaced, focusing on the selection of the training data and features based on their correlation with the main problems.

Additionally, we evaluated our method on more sequence-to-sequence models, out of which one was pretrained (specialized) in the domain of this study, to check if a specialization will help improve the performance. The models were fine-tuned on two well-represented languages in our dataset from the Indo-European family. Performance was measured on a zero-shot setting for most models using the underrepresented languages in our dataset. Additionally, for the zero-shot setting, we added a new language from the Uralic family, testing our model's robustness to languages from a different family than the ones it was fine-tuned on.

As a key result, we achieved a significant improvement in the F1 score for some of the most underrepresented classes in low-resource languages, increasing from 0% to 97%. This outcome highlights the effectiveness of our method in addressing the challenges of classification on highly imbalanced datasets.

The rest of this chapter is structured as follows: In the next section, Sect. 2, we review methodologies in multilingual and multitask learning, highlighting key advancements, our inspirations for the framework, and different architectures used to test it. Section 3 defines the research challenges, particularly concerning low-resource languages, and outlines the multitask classification scenarios and supporting data. In Sect. 4, we detail the preliminary steps and framework design, focusing on how instructional prompts enhance model performance across tasks and languages. The evaluation is presented in Sect. 5, where we analyze sequence-to-sequence models, decoder-only generation, and overall results across multiple configurations, including mT5 (specialized pretrained or not, with different sizes), ByT5 small, mBART, and GPT-2 Neo. These evaluations are conducted to demonstrate that our framework consistently improves performance regardless of the underlying model architecture/size and bridges the performance gap between smaller and larger models (answering the open question in [3] on the impact of model size and truncation on performance within the QdGF). Similarly, we examine how our method enhances pretrained models while also significantly improving non-pretrained versions, reducing the dependency on expensive pretraining. We highlight the framework's strong performance in low-resource languages and its ability to generalize effectively to unseen languages and the framework's effectiveness in boosting F1 score for the most underrepresented classes. We also perform an ablation study on the selection of individual questions composing our framework (Sect. 5.6). Finally, Sect. 6 summarizes our findings, discusses the implications of our research, and highlights improvements for low-resource languages and different model architectures.

2 Related Work

The concept of training a unified model to handle multiple tasks simultaneously, facilitating the sharing of knowledge and representations across tasks, has been extensively investigated in NLP applications. These range from intent detection and slot filling to more complex tasks such as joint entity and relation classification, co-reference resolution in scientific literature, and machine translation [4].

Liu et al. [5] introduces an innovative paradigm that redefines the traditional "Pre-train and Fine-tune" methodology commonly employed with large language models. Instead, it proposes the "Pre-train, Prompt, and Predict" approach, which leverages carefully crafted prompts to tap into the model's pre-existing knowledge without requiring extensive fine-tuning. This method relies on the use of predefined prompts to guide the model in generating relevant predictions directly, demonstrating how prompt engineering can streamline model usage and unlock its full potential by exploiting its training priors effectively.

In this direction, [6] introduces the unified generative framework (UGEN), which reformulates tasks as question-answering problems to handle joint intent detection and slot filling. During training, UGEN employs a set of five templated questions to extract key information from the input context, such as important keywords, thereby enhancing the model's ability to generalize. Interestingly, during evaluation, only two of the five questions are utilized, omitting the auxiliary questions used during training. This strategy demonstrates how targeted question templates can improve model performance by focusing on extracting critical features while reducing reliance on additional prompts during inference.

Generative models have revolutionized the landscape of natural language processing, unlocking unprecedented potential for leveraging the compositional abilities of language models through prompting. This approach spans multiple strategies, including fine-tuning, in-context learning, and augmenting models with reasoning skills or external tools, as demonstrated by works like [7] and [8]. By enhancing the interpretability and flexibility of models, these methods allow for tailored solutions to a wide range of complex tasks. For instance, [7] discusses how augmenting language models with external tools, such as calculators or databases, significantly enhances their reasoning and problem-solving capabilities. Similarly, [8] highlight the utility of task-specific prompts in guiding models to handle numerical reasoning challenges, bridging their gaps in this area.

An exemplary innovation in this domain is chain of thought prompting (CoT, [9]), which investigates the emergence of advanced reasoning capabilities in LLMs by employing a sequence of strategically crafted demonstrations. CoT prompting enables models to break down complex problems into structured, step-by-step logical progressions, enhancing their capacity to handle tasks that demand intricate reasoning, such as multistep arithmetic or logical inference. This approach not only improves the accuracy of predictions but also provides interpretable outputs that clarify the reasoning process behind the decisions.

Another example is the generative framework for multilingual, multitask classification proposed by us [3], emphasizing the importance of "asking the right questions" to uncover hidden task interactions. The approach leverages instructional prompts to reformulate tasks as questions, enabling the model to capture synergies across multiple tasks and languages. By integrating these prompts into a fine-tuning phase, the framework effectively improves classification performance, particularly in underrepresented languages and imbalanced datasets. We highlight the transformative role of prompt engineering in enhancing generative models for complex multilingual NLP applications.

mT5 [10], the generative model used in [3], is a versatile sequence-to-sequence model (the multilingual variant of T5 [11]) that reformulates all NLP tasks into a text-to-text format. This unified approach enables mT5 to excel in diverse tasks, including translation, summarization, and classification, by leveraging transfer learning and pretraining on the large-scale mC4 dataset.

A new model used to test our method, ByT5 [12], presents a token-free framework for sequence-to-sequence tasks. Unlike traditional models that rely on tokenized inputs, ByT5 processes raw byte-level data directly, eliminating the need for pre-tokenization. This approach simplifies multilingual NLP tasks by handling diverse scripts and rare languages uniformly, without relying on language-specific tokenizers. Despite the increased input length, ByT5 achieves competitive performance across a wide range of tasks, demonstrating the potential of byte-level models in reducing preprocessing complexity and improving language-agnostic capabilities.

Another model, mBART (multilingual bidirectional and auto-regressive transformer, [13]) is a sequence-to-sequence model designed for multilingual NLP tasks. It extends BART [14] by incorporating multilingual training data, using a denoising autoencoder objective to pretrain on diverse languages. mBART excels in translation and other multilingual tasks by learning shared representations across languages, making it especially effective in low-resource and cross-lingual scenarios.

Moving beyond sequence-to-sequence (Seq2Seq) toward decoder-based models for generation, GPT-Neo is an open-source implementation of *GPT-3*-like language models developed by EleutherAI [15]. GPT-Neo comes in various sizes, with the largest model containing 2.7 billion parameters. It was designed to replicate the capabilities of larger models, providing a competitive edge in terms of performance.

3 Problem Statement

Our target is to improve multitask, multilingual classification behavior using additional data/annotations available only during training, by using the MTL paradigm. To accomplish this, we used a private dataset in the cooking domain with additional annotations and some difficult main tasks that could benefit from this method. It consists of tens of thousands of scrapped web-based cooking recipes (including a wide range of public cooking websites), in six languages: English, German, Dutch, Italian, French, and Swedish. German and English are overrepresented in the labeled data, while Italian and Swedish are the most underrepresented. Each recipe has been annotated for three target tasks:

- Meat-centric (*M.*)—binary text label with "yes" or "no" classes, representing whether the main "focus" of the dish is (a loaf of) meat
- Surface aspect level (*S.l.*)—binary numeric label: classes "1" and "2," representing the cooked product's aspect/color
- Dehydration level (*D.l.*)—label with five numeric classes: from "1" to "5," representing the moisture reduction degree

Furthermore, in addition to the main classification tasks of *M.*, *S.l.*, and *D.l.*, each recipe has:

- A *title*
- Cooking *instructions*
- The required *oven settings*
- Several additional annotations related to the *dish type*, certain *ingredients*, *size* of the end dish, *thickness*—where applicable, etc.

For each recipe, the default text used by the model as input consists of the recipe's *title* and *instructions* extracted from the HTML content, as in [16]. To assess the cross-lingual capabilities of the methods investigated, we added recipes in a language from the Uralic family (Hungarian) as test data. It was obtained by translating recipes from test data in German, as in [17].

The recipe data used exhibit significant diversity in terms of structure and format, which assures that models may not simply learn the specifics of the cooking recipes themselves and suggests that the method can be generalized to other domains where additional/redundant data, extracted data from the text to classify, or even metadata available in a new dataset/domain are available.

4 Question-Driven Generative Framework (QdGF)

The central part of our approach is the question-driven generative framework (*QdGF*, introduced in [3]), which uses a template to generate different types of questions from redundant/auxiliary training data. Inspired from [6], our approach generates the response to the questions from auxiliary data available only during training and does not extract it from the target labels. To this extent, the model is fine-tuned in a multitask manner, but with a variable number of tasks, as some of the additional information is not available for all the recipes in the training data. The only tasks to be guaranteed for all recipes are the main tasks, transformed in the same manner as the other tasks, to questions. The intuition behind the approach is that, by providing the model with additional information during training, it can learn more effectively the relation between the textual information provided in the recipe's *title* and *instructions* and the three target labels.

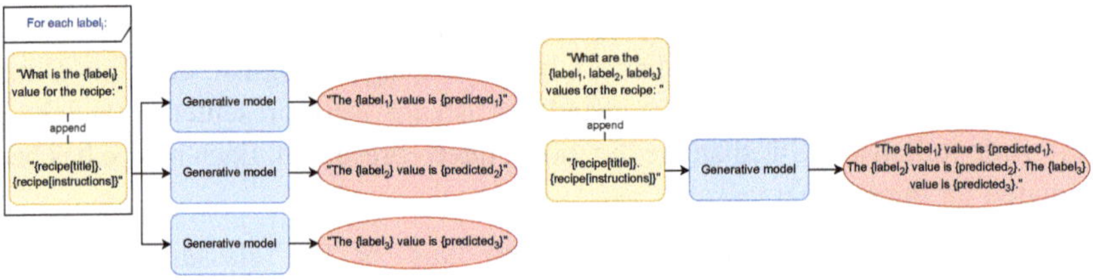

Fig. 1 Architecture of the baseline model (left) vs. architecture of the joint model (right) [3]

4.1 Preliminary Steps

Multitask Joint Learning Given the characteristics of the problem, we assume there is a certain amount of interaction between the three target labels. To exploit this, a preliminary step was to consider a *joint* model, fine-tuned on all three tasks simultaneously. The chosen order of generation is from the easiest task to the hardest task (as indicated by baseline, individual models: Lower accuracy obtained for the task means that the task is more difficult, as in [3]): $M. \rightarrow S.l. \rightarrow D.l.$ This way, the latter predicted labels should benefit from already having access to the labels predicted before. This is one of the reasons we employed a sequence-to-sequence generative model for our classification tasks, our main focus being the *S.l.* and *D.l.* tasks, as for these the baseline model seems to struggle more. The difference between the architecture of a baseline model and the joint model is illustrated in Fig. 1.

Oven Settings as Additional Learning Task Although not among the target tasks, we considered using the *oven settings* as an additional task in the joint model, only during fine-tuning, with the highest precedence compared to the target tasks:

OvenSettings → *M.* → *S.l.* → *D.l.*, as in [3]. The intuition was that the oven settings would have the greatest impact on the outcome (especially for *D.l.* and *S.l.* tasks, which should also be affected by whether the meat is the main content of the dish). The additional task of predicting the oven settings is formulated as predicting the triplet (T, t, P):

- Temperature setting (either in °C or °F)
- Time setting (in minutes)
- Cooking program

Some of the recipes in the available dataset contain multiple such triplets, either because they are multistep (in which case the triplets are joined with "AND") or because the recipe provides alternative cooking instructions (pairs of disjunctions—triplets joined by an "OR"), we applied the preprocessing procedure described in [3]. This model would be referred to as *OvenSettings*. The partial results for this specific task using a predefined *soft accuracy* are reported in Appendix.

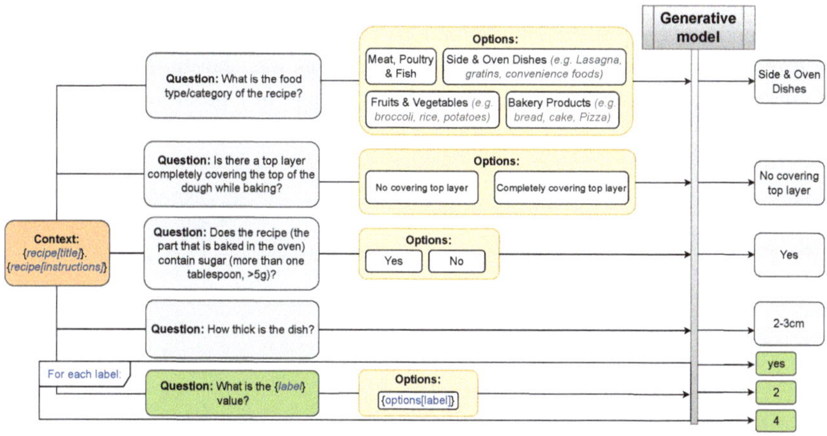

Fig. 2 $QdGF(\{Type, Top, Sugar, Thick\})$—question-driven generative framework with an example of question templates used only for training (gray boxes/first 4) and questions for the main tasks (*M.*, *S.l.*, and *D.l.*) used for testing (green box/last one), with the corresponding options, and their explanatory information (gray text in parentheses) [3]

4.2 Framework Description

QdGF relies on a series of instructional prompts to leverage additional training-time information, such as dish category, certain ingredients, end-product thickness, etc. The input prompts are constructed in the following format (see Fig. 2 for details):

$< s > Context : \{recipe[title]\}.\ \{recipe[instructions]\} < /s >\quad Question : \{Q_i\}?\ \ Options : \{options_i + explanations_i\} < /s >$ where $i \in S \mid \exists QdGF(S)$.

The number of questions/instructional prompts "*i*" used ranges between 4 and 8 per experiment (due to hardware resources constraint), without counting the three main problems (which add 1 or 3 more questions depending if they are independent or joint, respectively). All these questions represent a subset "*S*" from a set of 14 extracted questions initially. That is, we concatenate more information from the dataset (the recipe's title and instructions, together with several questions from the set of used questions, and their *answer options*, provided that the recipe is labeled with those annotations in the dataset) and use the resulting string as input to a generative model in the fine-tuning stage.

With these new prompts added, we try to stimulate the prior general knowledge of the model [18] and direct it to a greater focus on this downstream task composed of the three target problems (*M.*, *S.l.*, and *D.l.*). Several elements in the options list also contain additional explanations that help guide the model, but these were removed from the labeled option. Providing the options to choose from and some related explanations (where available) should ground the model to the current problem domain and prevent potential hallucinations [19].

A discussion on how the subset of used questions was chosen from the main set of questions is presented in Sect. 5.6, and an example of a possible instantiation of the framework ($S = \{Type, Top, Sugar, Thick\}$) is illustrated in Fig. 2.

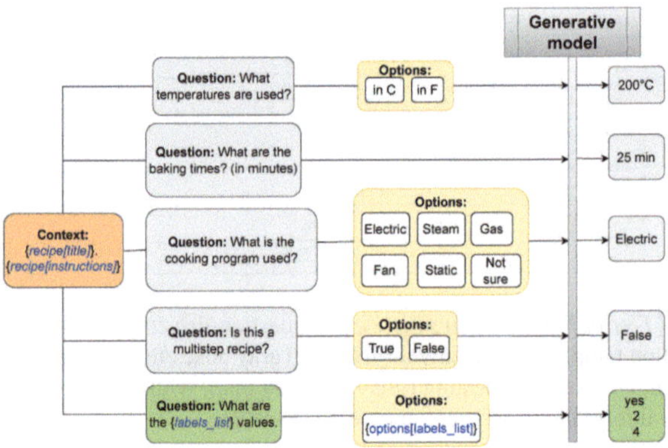

Fig. 3 $QdGF(\{Temperature, Time, Cooking\_progran, Multistep\})$—QdGF instantiated with the subset of OvenSettings questions [3]

QdGF + OvenSettings We transformed the triplet (T, t, P), from Sect. 4.1, into three more questions, one for each oven setting, to further exploit the previous approach and adapt it to this framework. We also added an additional question, to make a difference between multistep recipes (having more such triplets) and single-step recipes (with just one triplet). An example framework with this subset of questions is illustrated in Fig. 3. The total number of potential additional questions to choose from, including the oven settings-related questions, is now 18.

QdGF + Renaming Another potential issue we tried to overcome in this framework is related to the use of numbers to encode class labels, which bears either no, or a potentially wrong semantic meaning for language models [20]. To address it, we changed each class from its numeric counterpart to a textual description consisting of 1–3 words, which we considered to best capture the class meaning. We integrate these newly renamed labels directly into QdGF, with their respective explanations in the *Options* component.

All extracted questions are available in [3, Appendix B].

5 Experiments and Results

In this section, we evaluate our method through cross-lingual assessment, focusing on its ability to improve performance across diverse languages. Our analysis emphasizes generalization, assessing whether the approach effectively adapts to both high- and low-resource languages irrespective of the underlying architecture, tokenization method, or the number of parameters (size). By comparing results across different linguistic settings, we highlight the impact of our method on underrepresented classes and its robustness in multilingual tasks. For assessing the results, we considered the best experiments per problem out of the following (fine-tuned on De and En, zero-shot setting for Nl, Fr, It, Sv, Hu):

- **Joint**: All three problems at once, see Sect. 4.1.
- **OvenSettings**: Joint model with the additional fine-tuning task, see Sect. 4.1.
- **best QdGF**: Model with a question-driven generative framework (see Sect. 4.2), and only the best instantiation of the questions per model and per language is reported in the final results. Results from other instantiations are discussed in Sect. 5.6.

The overall results were considered per language, and for all languages combined using the weighted accuracy (with respect to the number of samples of each language). Since Hungarian test data was obtained from German test data by translation, we did not include it in the computation of the final weighted accuracy for all languages.

The baseline used to compare our results is one model per problem (*M.* or *S.l.* or *D.l.*) fine-tuned on De and En (dubbed *Baseline*) with a simple instructive prompt and the default input (*title+instructions*, see Sect. 3). We report the average result of multiple runs (with standard deviations up to 0.1%) for every mentioned experiment. Since some of the tasks/questions need text generation, we had to use generative models, mainly Seq2Seq models (as our approach works better for this type of architectures): *mT5 small*, *mT5 base*, *mT5 large*, *PT mT5 base*, and *PT mT5 large* (the "Specialized" pretrained versions, see 5.1.1), *ByT5 small*, *mBART*, and one decoder-only model *GPT-2 Neo*.

Table 1 Testing dataset's imbalance ratios per language (first column, last row), imbalance ratios per language and per task (with two, two, and five classes, respectively), and imbalance ratios per task (last row, last three columns)

Lang.	M.	S.l.	D.l.
German	12:1	2:1	12:6:4:2:1
English	9:1	2:1	10:6:5:1:1
Dutch	7:1	1:1	10:8:8:3:1
French	12:1	2:1	4:2:2:1:1
Italian	17:1	2:1	7:4:4:4:1
Swedish	13:1	2:1	7:6:4:3:1
Hungarian	12:1	2:1	12:6:4:2:1
15:7:4:3:3:1:15	10:1	2:1	8:5:3:1:1

The dataset used contains approximately 52000 multilingual recipes labeled with the three target tasks and several additional annotations (few of them being sparse), performed by human experts (see Sect. 3). It is highly imbalanced both respective to the languages and the classes used (see Table 1). Imbalances per language were reported only for the testing dataset, as the training and validation datasets only contain De and En languages, for them the ratio is $38 : 16$.

The train-validation-test split is 0.64–0.20–0.16 for German and English (the fine-tuning languages), and it is stratified according to the ratio of classes from Table 1. We fine-tuned only with the high-resource languages. The other languages are used in a zero-shot setting. We added Hungarian also for the zero-shot setting (translated from the German test data, as mentioned in Sect. 3).

5.1 Seq2Seq

Since our sequence-to-sequence method works best with encoder-decoder transformer architectures [21], we used several such models, starting with *mT5* [10], a multilingual encoder-decoder transformer based on the T5 model (text-to-text transfer transformer [11]). We used several sizes of the *mT5* model, *small*, *base*, and *large*. Then, we used *byT5* [12], a model that uses the same architecture as before, but with a byte-level encoding. This type of encoding increases the necessary space, and thus we were able to use only the *small* size within our resources. The last Seq2Seq model that we used is *mbart-large-50* [13].

Through manual hyperparameter tuning, we found the following hyperparameter values to produce the best results (and were used further for all Seq2Seq models): *Batch size: 8*, *learning rate:* $3e-4$, *weight decay:* 0.1, *no warm-up* with optimizer *AdaFactor* (the original optimizer used for the T5 model [22] which produced better results than AdamW [23]) and its scheduler: *AdafactorSchedule*.

5.1.1 Specialization Pretraining

We used the mT5 model pretrained by Google (with sizes *base*[1] and *large*[2]) on mC4 and a further "Specialized" model on culinary data (further pretrained using span-masked language modeling on a multilingual dataset of approximately 490 000 unlabeled recipes for the model already pretrained on mC4 dataset, [3]).

5.2 Decoder-Only Generation

We extend QdGF from the Seq2Seq paradigm to a single-stage, decoder-only generation process in order to test its compatibility with different architectures. For this, we used EleutherAI's *gpt-neo-1.3B* model [15].

To structure input sequences when applying QdGF's multitask learning with a decoder-only model like GPT-2 Neo, we added a separator token, $< sep >$, as a special token which is incorporated during tokenization. Since these models process inputs autoregressively, adding a $< sep >$ token explicitly separates different components of the input, such as the context and question from the expected answer. This helps the model distinguish between segments while maintaining coherence in generation. A typical input format is $< s >$ $Context : \{recipe[title]\}.\ \{recipe[instructions]\} < /s >$ $Question : \{Q_i\}?$ $Options : \{options_i + explanations_i\} < /s >$ $Answer :< sep >$ allowing the model to recognize boundaries and generate appropriate completions. To ensure the model learns only from the answer and not the context or question, we mask non-label tokens with the help of the separator token. This way, GPT-2 Neo focuses solely on predicting the answer, improving generalization and preventing overfitting to the input structure.

Despite this masking technique, GPT-2 Neo, being a causal language model, still generates answers that are less structured compared to a Seq2Seq model like *mT5*. This happens because GPT-2 Neo predicts tokens autoregressively, relying solely on past context without an explicit separation between input and output spaces. Unlike encoder-decoder architectures that condition the entire response on a well-defined input representation, GPT-2 Neo continues generating text based on probabilities, sometimes diverging from structured answers. To handle this, we extract the first generated word that matches one of our predefined class labels, ensuring that the model's output aligns with the expected categorical format and for ease of computation of the final results. The predicted output was converted into lower case.

For fine-tuning, the final optimizer we used is *AdamW* [23], which improves upon *Adam* by decoupling weight decay from gradient updates, leading to better generalization. The learning rate follows a *cosine schedule* with *warm-up*, where the first 5% of total training steps are used for warm-up to stabilize early training. The peak *learning rate* is set to $3e-5$, a commonly used value for transformer fine-tuning,

[1] https://huggingface.co/google/mt5-base

[2] https://huggingface.co/google/mt5-large

and a *weight decay* of 0.01 is applied to mitigate overfitting by discouraging large parameter updates. The *batch size* had to be set to 1, due to memory constraints caused by the model's large number of parameters. Since this significantly affects training dynamics, it was essential to compensate using *gradient accumulation* to simulate larger batch sizes (instead of updating weights after every sample, accumulate gradients over multiple steps before performing an update), stabilizing training [24].

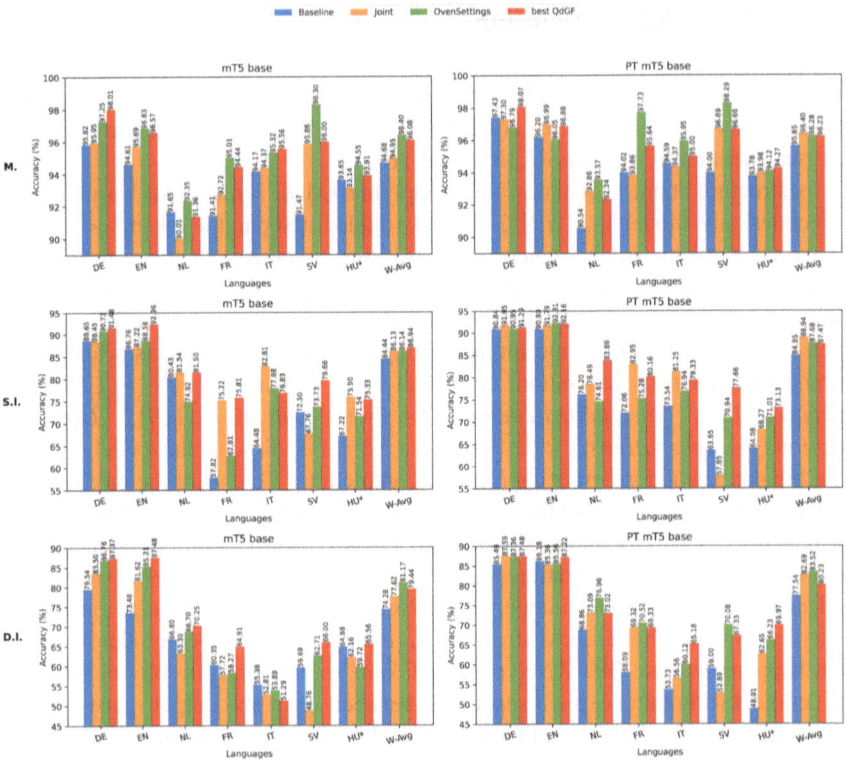

Fig. 4 Results per problem (each row) for *mT5 base* model and the specialized pretrained *mT5 base*. * - language not included in the final weighted accuracy

5.3 Overall Results

5.3.1 *mT5 Base* and Specialized *PT mT5 Base*

The results for *mT5 base* models as seen in Fig. 4, highlight the importance of using additional data in improving generalization and overall performance across different languages. Across the evaluated languages, tasks involving additional data, such as OvenSettings and QdGF, consistently outperform the baselines, with particularly strong improvements in low-resource languages such as Hungarian and Swedish.

Across all tested configurations, the specialized models outperform by a small margin their nonspecialized counterparts, demonstrating that task-specific pretraining indeed brings some benefits. However, there are cases where the additional pretraining seemed to have negative impact on the low-resource languages, such as Swedish and Hungarian for the *S.l.* problem (Swedish being low represented also on the pretraining dataset, while Hungarian being absent entirely). QdGF seems to be the least affected though, keeping the lead in this case, proving that its generalization ability can surpass that lack of representation of certain languages in the pretraining dataset.

The greatest absolute improvements in accuracy from the baseline obtained by QdGF for the *base* model are 4.53% in Swedish for *M.*, 17.99% in French for *S.l.* and 14.02% in English for *D.l.*, while for the specialized *base* are 2.66% in Swedish for *M.*, 14.01% in Swedish also for *S.l.*, and 21.06% in Hungarian for *D.l.* The results in zero-shot Hungarian suggest that QdGF is able to generalize well even in new and different languages.

QdGF can reduce the need of specializing the model (a very expensive task since it requires pretraining again and impossible where data is scarce), as there are absolute improvements in accuracy from QdGF applied to a nonspecialized model over a specialized baseline of up to 2% in Swedish for *M.*, 16.01% in Swedish for *S.l.*, and 16.65% in Hungarian for *D.l.*

The maximum number of tokens was set to 512 for both models, approximately 500/55000 recipes truncated.

5.3.2 *mT5 Large* and Specialized *PT mT5 Large*

The evaluation of QdGF-enhanced models in comparison to standard and specialized pretrained *(PT) mT5 large* models reveals significant improvements in accuracy across different languages, as seen in Fig. 5. The results consistently demonstrate that QdGF enhances generalization, with the most notable gains observed in low-resource languages such as Hungarian and Swedish.

Pretraining, which is a very expensive task with respect to both its complexity (time and memory) and the amount of data it requires, improves overall accuracy, particularly in mid-to-low-resource languages such as Dutch and Swedish. However, pretraining alone does not fully bridge the gap between languages, while QdGF extends these benefits by enabling more flexible contextual understanding across

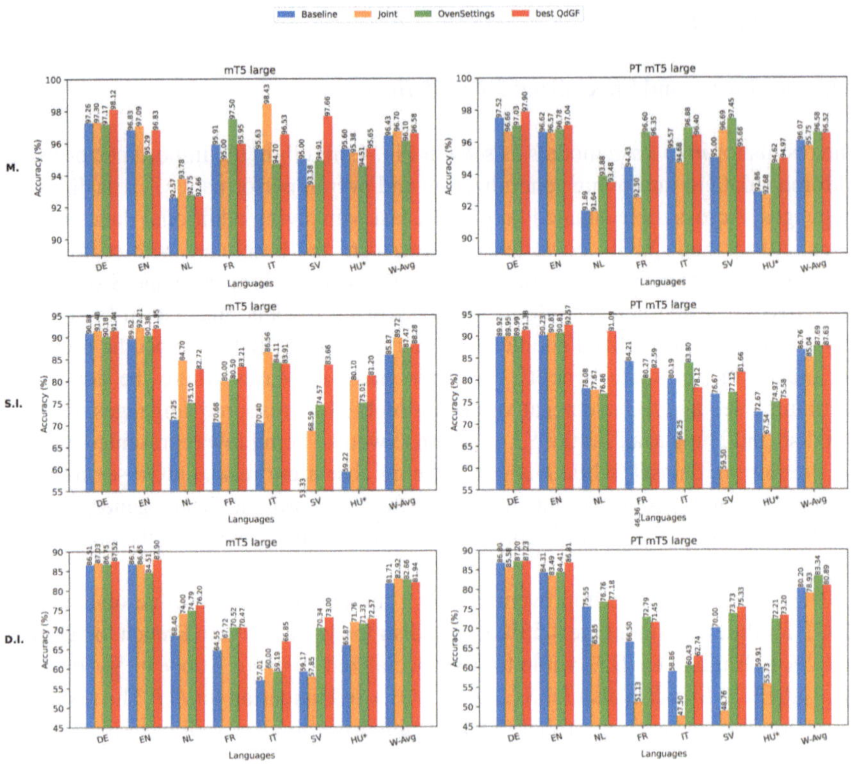

Fig. 5 Results per problem (each row) for *mT5 large* model and the specialized pretrained *mT5 large*. * - language not included in the final weighted accuracy

multiple language families. We can see that the results of QdGF are similar between the specialized and nonspecialized model, sometimes even lower by a small margin in the specialized model (this can happen due to the combined complexity of additional pretraining and QdGF for the limited hardware resources available). On the other hand, the *S.I.* and *D.I.* results for zero-shot languages are much better for the baseline with the specialized model, but still under the base model with QdGF, suggesting that a good QdGF can replace the need of specialization pretraining, bringing significantly better results with lower resources and diminished complexity. Moreover, the fact that specialization pretraining in the cooking domain did not help significantly suggests that the method can be generalized to other domains, by instantiating a QdGF with questions obtained from additional/redundant data.

The absolute improvements in accuracy from the baseline obtained by QdGF for the *large* model are 2.66% for *M.*, 30.33% for *S.I.*, and 13.83% for *D.I.*, all in the Swedish language; and for the specialized *large* QdGF are 2.11% in Hungarian for *M.*, 13.01% in Dutch for *S.I.*, and 13.29% in Hungarian for *D.I.*, while the greatest absolute improvements in accuracy obtained by QdGF for the nonspecialized *large* model over the baseline specialized model are 2.79% for *M.*, 8.53% for

S.l., and 12.66% for *D.l.*, all in the Hungarian language, emphasizing that QdGF can eliminate the necessity for specialized pretraining while still achieving strong performance in low-resource languages. This highlights its ability to enhance cross-lingual generalization, making it a more scalable and adaptable solution for multilingual tasks.

For the *large* models, we had to set the maximum number of tokens to 320, approx. 2600 recipes being affected, and for the more complex approaches (QdGF), we had to set the maximum number of tokens to 254, 5100+ recipes being affected, thus the small reduction in performance for QdGF from nonspecialized to specialized (more on this in Sect. 5.5). Despite this additional truncation, QdGF for specialized models still performed best compared to the other experiments that had an unfair advantage as for them truncation remained the same as the nonspecialized model. Once again, QdGF generalization ability proves to counter the extra truncation needed, keeping its performance in the top choices.

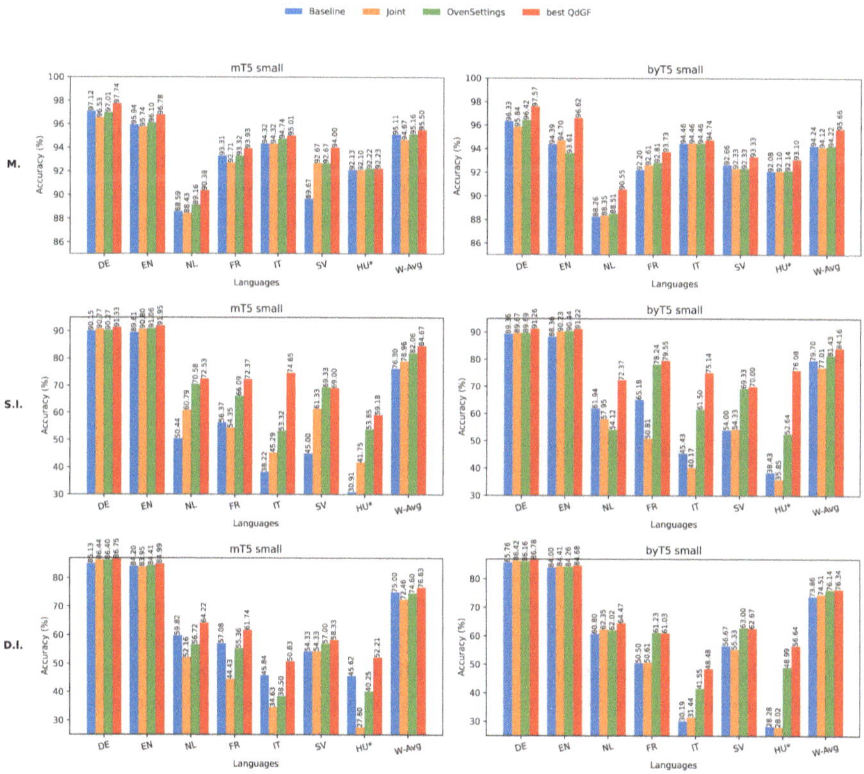

Fig. 6 Results per problem (each row) for *mT5 small* and *ByT5 small*. * - language not included in the final weighted accuracy

5.3.3 *mT5 Small* and *ByT5 Small*

All previous models used standard T5 tokenization methods, while ByT5 uses byte-level tokenization [12]. This new approach, while innovative, uses more memory, so the only size able to fit within our resources was *small*. The maximum number of tokens was set to 512 for mT5, while for ByT5 was set on average ~3 times more. We used an *mT5 small* model to prove QdGF can still generalize irrespective of the tokenization method applied to the same architecture. The results can be seen in Fig. 6.

For the first two problems (the binary classification ones), ByT5 seems to perform much better in low-resource languages than mT5, for the baseline experiment, proving that eliminating the need of standard tokenization improves generalization over different languages. However, QdGF seems to lower this gap between the two tokenization methods, making them similar in performance, and both better than their respective baselines (and than the other baseline too, for each tokenization method). In both cases, not only QdGF outperformed the baseline every single time (joint and OvenSettings methods too), but it brought massive improvements over them, bringing the *small* sizes closer to their *base* or even *large* counterparts.

The most notable absolute improvements in accuracy obtained by QdGF over the baseline for the mT5 model are 4.33% in Swedish for *M.*, 36.43% in Italian for *S.l.* (and other 22%+ in Dutch, Swedish, and Hungarian), and 6.59% in Hungarian for *D.l.*, while with ByT5: 2.29% in Dutch for *M.*, 37.65% in Hungarian, and 29.71% in Italian for *S.l.* and 28.36% in Hungarian for *D.l.* The zero-shot results suggest that QdGF is able to generalize well even with a byte-level tokenization method applied.

Hallucinations occurred extremely rare for the T5-based models. A rare occurrence would be the case of ByT5. When testing in Hungarian, the model predicted rarely “*ne*” instead of “*no*” for *M.* problem. These were considered wrong answers.

5.3.4 *mBART* and *GPT-2 Neo*

The results from mBART and GPT-2 Neo, from Fig. 7, suggest once again that QdGF enhances model generalization, leading to better cross-lingual performance, particularly in low-resource languages such as Hungarian and Swedish. Additionally, mBART consistently outperforms GPT-2 Neo for the *M.* problem, while for the *S.l.* happens the opposite and for *D.l.* GPT-2 Neo leads in low-resource languages only, but these gaps narrow down significantly when QdGF is applied. Furthermore, mBART consistently outperforms GPT-2 Neo in baselines. However, after applying QdGF, GPT-2 Neo reaches competitive levels and, in some cases, outperforms mBART as QdGF enables GPT-2 Neo to generalize significantly better in lower-resource settings. The greatest improvements are observed in GPT-2 Neo because it initially lacks strong multilingual generalization, which QdGF helps overcome.

While the joint method performs well on *M.* problem in all languages and also in well-represented languages for *S.l.* problem for both models, it struggles in low-represented languages for *S.l.* problem for both models and in all languages for *D.l.*

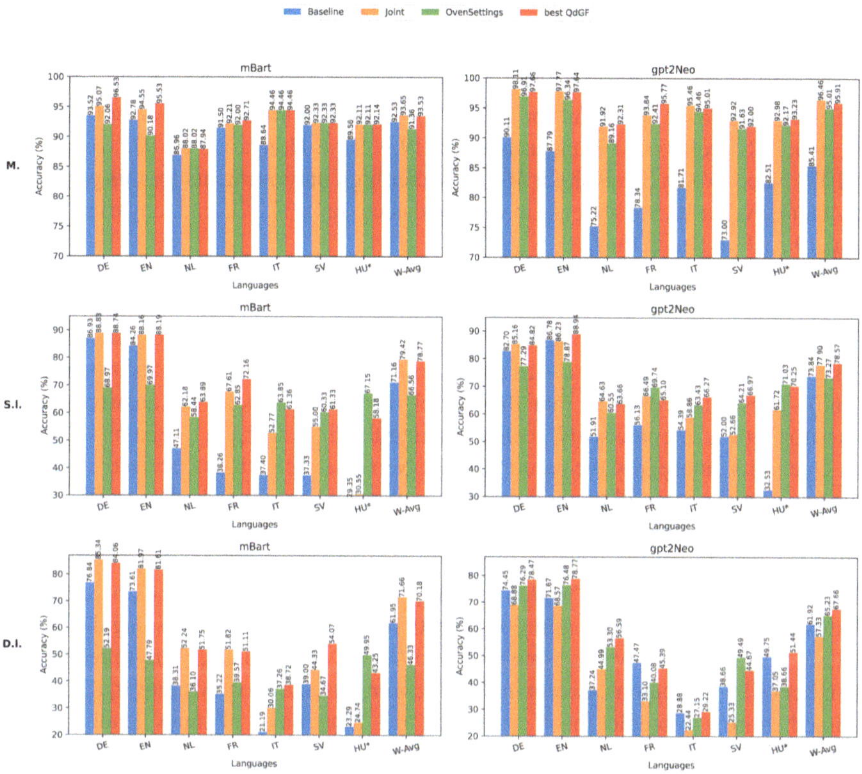

Fig. 7 Results per problem (each row) for *mBART* and *GPT-2 Neo*. * - language not included in the final weighted accuracy

problem with GPT-2 Neo, where the OvenSettings method does exactly the opposite. Meanwhile, QdGF's behavior remains constant as the best method, or very close in terms of performance to the best method, being either the first or the second best, for both models. This highlights its generalization ability, not only across languages, but also across different problems and architectures, making it the best choice in terms of performance for every model.

The OvenSettings method achieves high results for GPT-2 Neo and other architectures, except for mBART. mBART seems to struggle to predict the three main problems using this method, where it has to predict the (T, t, P) triplet first. A deeper analysis on this triplet alone shows that mBART struggles to predict its elements, falling behind every other model for T (Temperature) and t (time) categories, with extremely low results, even for our predefined soft accuracy (as can be seen in Appendix). This shows that proficiency at the other tasks in a multitask paradigm matters for every task's overall performance, making it important to select carefully the tasks (or questions in the case of QdGF; the questioning strategy is talked about in Sect. 5.6).

The maximum number of tokens was set to 512 for mBART and to 384 for GPT-2 Neo, due to its larger size.

Significant absolute improvements in accuracy of QdGF over the baseline for mBART are 5.82% in Italian for *M.*, 33.9% in French for *S.l.* (and other ~ 24%+ in Italian, Swedish, and Hungarian), and 19.96% in Hungarian for *D.l.*, while for GPT-2 Neo: 19% in Swedish for *M.* (and other high improvements for this particular problem, even a 10.5% increase in weighted accuracy), 37.72% in Hungarian for *S.l.* and 19.35% in Dutch for *D.l.*

Out of all Seq2Seq models, mBART had the most "hallucinations," with the QdGF method that used the questions obtained from the OventSettings task (Sect. 4.2) in Hungarian, by predicting the temperature and time during inference instead of the asked questions (*M.*, *S.l.*, *D.l.*—as these are the only used questions/tasks on testing time). These indeed show that the model intrinsically looks for answers to the questions it was prepared with, before making the final decision on the main tasks.

Due to its autoregressive nature, GPT-2 Neo's hallucinations occurred across all languages and across all methods, where it mistakenly predicted other numbers that do not belong in classes "1" and "2" for *S.l.* and in classes "1" to "5" for *D.l.* (like "8" for example). As numerical comprehension is not one of its strengths [8], this led us to believe that *QdGF with Renaming* might help (Sect. 4.2), but, on contrary, it missed the named classes entirely or partially most of the time (mixing the names within the classes, or even between *S.l.* and *D.l.*). QdGF did help here to reduce the number of hallucinations, except in the *renaming* instantiation. Moreover, the model missed entirely to predict the *M.* problem ("yes"/"no") a lot of times, leading to the lowest results on this problem across all models. QdGF barely helped in this case.

5.4 Closing Another Gap: Boosting F1 Score for Underrepresented Classes in Low-Resource Languages

The QdGF approaches also helped to recognize the lowest represented classes, as the dataset is highly imbalanced, especially for *D.l.* task (Table 1, last column). For this, we measured the F1 score for the two most underrepresented classes in this problem, and the comparison can be seen in Table 2. A reason for the performance discrepancies in Table 2—*baseline row* between the second most underrepresented class (with an F1 score of 0 for underrepresented languages) and the most underrepresented class (with F1 score >60)—might be the model overfits on De and En (thus a higher F1 score for that class for De, En) and is unable to generalize to the other languages. We can see that employing QdGF reduces the overfitting significantly. We tested here on specialized *PT mT5* for QdGF and baseline also to measure only the improvements of QdGF and not the "specialization."

Table 2 Specialized *PT mT5* for *D.l.* problem—F1 scores for the two most underrepresented classes out of the five classes (2ndUnderrepClass/mostUnderrepClass). blue = Best *base*, teal = Best *large*

Method	Size	De	En	Nl	Fr	It	Sv
Baseline	Base	90 / 71	**94** / 89	0 / 87	0 / 77	0 / 60	25 / 69
	Large	89 / 81	92 / 86	18 / 87	0 / 78	0 / **66**	17 / 67
QdGF	Base	**91 / 86**	84 / 89	**90** / 87	**55 / 83**	**70 / 67**	**94 / 76**
	Large	**91 / 88**	**98 / 88**	**83 / 88**	**49 / 85**	**66** / 59	**97 / 75**

5.5 Size Does Matter, But Not for QdGF

To address some limitations and answer the open question in [3], we reduced the maximum number of tokens for a *PT mT5 base* model to be the same as its *PT mT5 large* counterpart (320 and 254, instead of 512) for the proposed methods. Truncation affected performance mostly in the specialized pretrained version of *mT5 large* due to its additional complexity, and mostly on certain QdGF instantiations, so in this analysis we focused only on specialized models, with those affected QdGF instantiations (not the best QdGFs).

As we can see from the results in Fig. 8, the number of parameters seems to affect performance for the OvenSettings task, as for all problems, *large* performs significantly better. Since this task requires to jointly predict extra information before giving the final answer, this is reasonable as more memory is needed to predict four different tasks at the same time. On the other hand, while, for the joint method, similar behavior is observed for the well-represented languages, for the low-resource ones, *base* performs much better in the more difficult problems (*S.l.* and *D.l.*), suggesting that some overfitting took place for *large*.

For QdGFs using numeric classes, the two sizes seem to be nearly equal in performance with no winner for the *M.* and *S.l.* problems. In contrast, for *D.l.*, the *base* size seems to surpass *large*, especially in underrepresented languages, hinting that exploitation of hidden correlations using the right questions is indeed more important than the size and number of parameters of the model. This framework might reduce the need for memorization, where large language models tend to be better [25].

For QdGF with renaming, *large* is significantly better, especially for underrepresented languages and for the difficult problems (the former problems with numeric classes and the only two actually affected by renaming). This might be due to the model having significantly more parameters, allowing it to memorize patterns better. For named classes, the large model can better capture rich contextual embeddings, improving classification accuracy, while for numeric classes, a large model might overfit to noise instead of learning robust number patterns, leading to worse generalization. Numeric patterns are simpler, so a smaller model is sufficient to learn them. *Base* handles this well because it learns general numeric structures, whereas *large* might overspecialize on token distributions seen during pretraining.

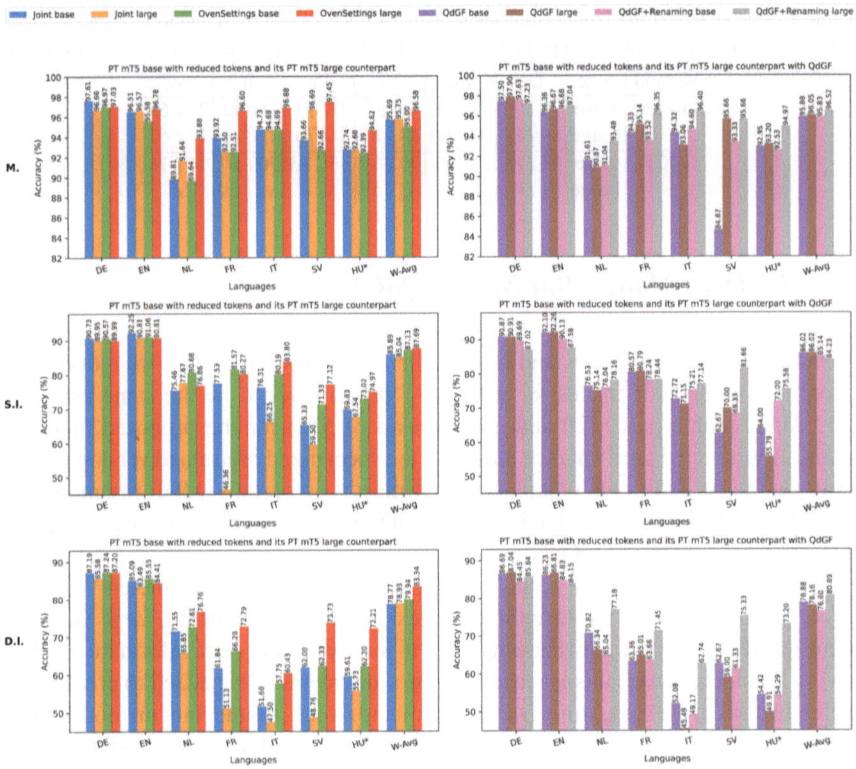

Fig. 8 Results per problem (each row) for specialized *PT mT5 base* with reduced maximum number of tokens to match its specialized *PT mT5 large* counterpart. Not the best QdGF instantiations are selected here, only where *large* seemed affected by additional truncation. * - language not included in the final weighted accuracy

5.6 Ablation Study

The set of 18 available questions (see Sect. 4.2) results in 2^{18} possibilities of instantiating QdGF:

$$|\{QdGF(S) \mid S \subseteq Q\}| = 2^{|Q|}, \tag{1}$$

where:

- S represents any subset of Q.
- $QdGF(S)$ represents an instantiated version of the question-driven generative framework with the subset of questions S.
- $|Q| = 18$ the total number of extracted questions, available in [3, Appendix B].

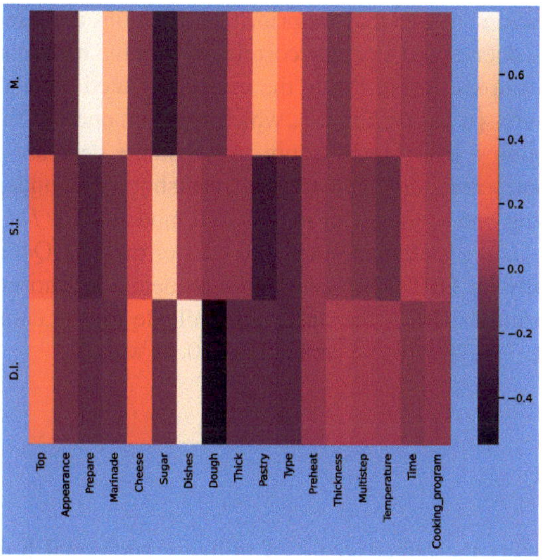

Fig. 9 Relevant correlations between *M.*, *S.l.*, *D.l.*, and answers of selected questions (Pearson correlation)

To select the subset of questions, we avoid the most sparse questions in the dataset and ensure a strong correlation between the selected questions and our three main classification problems. The optimal subset of questions Q' is chosen as

$$Q' = \arg\max_{Q^* \subseteq Q} \left(\sum_{q \in Q^*} \mathbb{1}[\text{density}(q) > \tau] + \lambda \sum_{q \in Q^*} \sum_{p \in P} \text{corr}(q, p) \right) \tag{2}$$

$$\text{subject to } |P| < k_{\min} \leq |Q^*| \leq k_{\max}, \tag{3}$$

where:

- Q is the full set of available questions.
- Q' is the selected subset of questions that maximize relevance.
- $\mathbb{1}[\text{density}(q) > \tau]$ is an indicator function ensuring only non-sparse questions are included, based on a threshold τ.
- $\text{corr}(q, p)$ represents the Pearson correlation between question q and problem p, ensuring the selection of the most relevant questions.
- P is the set of the three main classification problems (*M.*, *S.l.*, *D.l.*).
- λ is a weighting factor balancing the importance of density vs. correlation.
- $k_{\min}$ and $k_{\max}$ are the lower and upper bounds for the number of selected questions. In our case, $k_{\min} = 4$ to have more questions than the number of problems (>3), so we can assume at least one question being useful per problem, and $k_{\max} = 8$ as constrained by learning capacity of the relatively small language models we were able to fit within our memory.

This maximization ensures that we prioritize questions that are both sufficiently represented in the dataset and strongly correlated with the main classification tasks. For example, we can see in Fig. 9 that the most correlated types of questions with the *S.l.* problem are $\{Cheese, Sugar\}$, which are contained in the best model for this problem in almost every architecture. An absolute decrease of 5.22% occurs in weighted accuracy zero-shot, and 15.56% zero-shot for an underrepresented language, if discarding one of them (for mT5-based models). For *D.l.*, the most correlated questions ($\{Dough, Dishes\}$) are also part of the best QdGF model for this problem (especially for GPT-2 Neo and ByT5 models). All chosen subsets contain the $\{Type\}$ question as this is the only one available for the entire dataset, and it correlates approx. -0.20 with *S.l.* and *D.l.* and 0.32 with *M*.

6 Conclusion

Our study highlights the effectiveness of utilizing additional training data as instructional prompts in a multilingual, multitask classification problem through the introduction of QdGF (question-driven generative framework). The additional questions provided during training force the model to more effectively learn the relation between the textual information provided in the recipe title and description and the target labels. We evaluate our approach on a multilingual recipe dataset with three target classification labels, characterized by both class and language imbalance, using multiple generative models. We focus specifically on assessing the cross-lingual capabilities of the approach, as well as its effectiveness for both decoder-only and encoder-decoder architectures and for both their standard tokenization or byte-level tokenization. Additionally, we study the effect of model size and in-domain pretraining, where QdGF closes the performance gap between each, bringing significantly better results with lower resources and diminished complexity. The best instantiation of the framework is obtained for the subset of questions that optimally balances question density and task correlation, producing absolute improvements in the weighted multilingual accuracy in all three classification problems: +10.5% for *M.* using GPT-2 Neo, +8.37% for *S.l.* using mT5 small, +8.23% for *D.l.* using mBART over each individual baseline. Moreover, our findings emphasize the critical role of strategic training data selection and additional questions, particularly for underrepresented languages (where we obtained absolute improvements over the baseline of 19% in Swedish for *M.* on GPT-2 Neo, 37.65% for *S.l.*, and 28.36% for *D.l.* both in Hungarian on ByT5) and imbalanced datasets (where we obtained an absolute increase of 90 in F1 score for the most underrepresented class of *D.l.* over the baseline using mT5 base in Dutch).

Appendix

In order to test how well each model extracts the additional task used to jointly help our three main tasks, we defined a *soft accuracy* for the predicted triplet (T, t, P).

Let:

- P_i be the *predicted set* for the ith sample.
- T_i be the *true set* for the ith sample.
- N be the *total number of samples*.

For each category, we define a *partial match function*:

$$M_k(i) = \begin{cases} 1, & \text{if } P_i^k \cap T_i^k \neq \emptyset \\ 0, & \text{otherwise} \end{cases},$$

where $k \in (T, t, P)$ corresponds to one of the three categories.

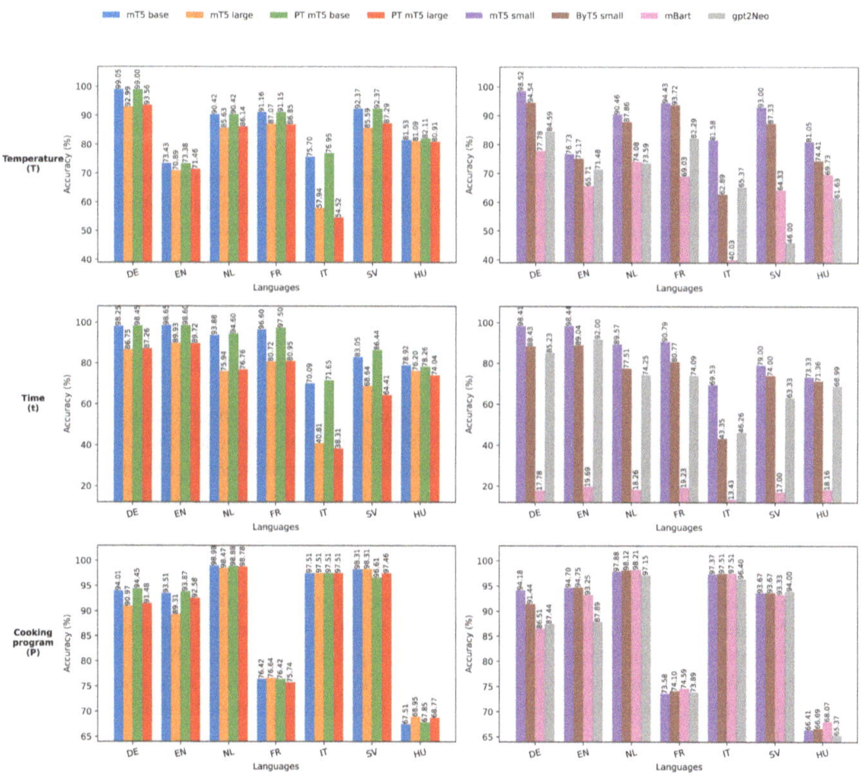

Fig. 10 Results using a *soft accuracy* per oven setting T, t, P (each row) for each architecture used in this study

The *soft accuracy* for each category is then computed as

$$A_k = \frac{1}{N} \sum_{i=1}^{N} M_k(i) \times 100,$$

where:

- A_k represents the *accuracy for category k*.
- $M_k(i)$ checks if *any predicted value for sample i matches a ground-truth value.*

Thus, the final soft accuracy results in: $\{A_T, A_t, A_P\}$. Results for each architecture can be seen in Fig. 10.

References

1. Li, Q., Peng, H., Li, J., Xia, C., Yang, R., Sun, L., Yu, P. S. & He, L. (2020). A survey on text classification: From shallow to deep learning. https://arxiv.org/abs/2008.00364
2. Hupkes, D., Giulianelli, M., Dankers, V., Artetxe, M., Elazar, Y., Pimentel, T., Christodoulopoulos, C., Lasri, K., Saphra, N., Sinclair, A., Ulmer, D., Schottmann, F., Batsuren, K., Sun, K., Sinha, K., Khalatbari, L., Ryskina, M., Frieske, R., Cotterell, R., & Jin, Z. (2023). State-of-the-art generalisation research in NLP: A taxonomy and review. https://arxiv.org/abs/2210.03050
3. Toma, S. A., Lemnaru, C., Negru, V. A., & Potolea, R. (2024). Asking the right questions: Exploiting hidden interactions in a generative framework for multilingual, multitask classification. In *Proceedings of the 7th International Conference on Natural Language and Speech Processing (ICNLSP 2024)* (pp. 251–266).
4. Chen, S., Zhang, Y., & Yang, Q. (2021). Multi-task learning in natural language processing: An overview. https://arxiv.org/abs/2109.09138
5. Liu, P., Yuan, W., Fu, J., Jiang, Z., Hayashi, H., & Neubig, G. (2021). Pre-train, prompt, and predict: A systematic survey of prompting methods in natural language processing. https://arxiv.org/abs/2107.13586
6. Wu, Y., Wang, H., Zhang, D., Chen, G., & Zhang, H. (2022). Incorporating instructional prompts into a unified generative framework for joint multiple intent detection and slot filling. In *Proceedings of the 29th International Conference on Computational Linguistics* (pp. 7203–7208). International Committee on Computational Linguistics. https://aclanthology.org/2022.coling-1.631
7. Mialon, G., Dessí, R., Lomeli, M., Nalmpantis, C., Pasunuru, R., Raileanu, R., Rozière, B., Schick, T., Dwivedi-Yu, J., Celikyilmaz, A., Grave, E., LeCun, Y., & Scialom, T. (2023). Augmented language models: A survey. https://arxiv.org/abs/2302.07842
8. Al-Negheimish, H., Madhyastha, P., & Russo, A. (2021). Numerical reasoning in machine reading comprehension tasks: Are we there yet? https://arxiv.org/abs/2109.08207
9. Wei, J., Wang, X., Schuurmans, D., Bosma, M., Xia, F., Chi, E., Le, Q., & Zhou, D. (2022). Chain-of-thought prompting elicits reasoning in large language models. https://arxiv.org/abs/2201.11903
10. Xue, L., Constant, N., Roberts, A., Kale, M., Al-Rfou, R., Siddhant, A., Barua, A., & Raffel, C. (2020). mt5: A massively multilingual pre-trained text-to-text transformer. https://arxiv.org/abs/2010.11934
11. Raffel, C., Shazeer, N., Roberts, A., Lee, K., Narang, S., Matena, M., Zhou, Y., Li, W., & Liu, P.J. (2020). Exploring the limits of transfer learning with a unified text-to-text transformer. https://arxiv.org/abs/1910.10683

12. Xue, L., Barua, A., Constant, N., Al-Rfou, R., Narang, S., Kale, M., Roberts, A., & Raffel, C. (2022). ByT5: Towards a token-free future with pre-trained byte-to-byte models. *Transactions of the Association for Computational Linguistics, 10*, 291–306. https://aclanthology.org/2022.tacl-1.17
13. Liu, Y., Gu, J., Goyal, N., Li, X., Edunov, S., Ghazvininejad, M., Lewis, M., & Zettlemoyer, L. (2020). Multilingual denoising pre-training for neural machine translation. In Transactions of the Association for Computational Linguistics, 8 (pp 726–742)
14. Lewis, M., Liu, Y., Goyal, N., Ghazvininejad, M., Mohamed, A., Levy, O., Stoyanov, V., & Zettlemoyer, L. (2019). BART: Denoising sequence-to-sequence pre-training for natural language generation, translation, and comprehension. In D. Jurafsky, J. Chai, N. Schluter & J. Tetreault (Eds.), *Proceedings of the 58th Annual Meeting of the Association for Computational Linguistics* (pp. 7871–7880). Association for Computational Linguistics. https://aclanthology.org/2020.acl-main.703
15. Black, S., Gao, L., Wang, P., Leahy, C., & Biderman, S. (2021). GPT-Neo: Large scale autoregressive language modeling with mesh-tensorflow. https://doi.org/10.5281/zenodo.5297715
16. Migea, A.-M., Negru, V.-A., Toma, S.-A., Lemnaru, C., & Potolea, R. (2024). Cook smarter not harder: Enhancing learning capacity in smart ovens with supplementary data. In *2024 IEEE 20th International Conference on Intelligent Computer Communication and Processing (ICCP)* (pp. 1–8). IEEE.
17. Veres, N., Negru, V.-A., Toma, S.-A., Lemnaru, C., & Potolea, R. (2024). Overcoming language barriers in classification through translation-augmented data. In *2024 IEEE 20th International Conference on Intelligent Computer Communication and Processing (ICCP)* (pp. 1–07). IEEE.
18. Han, X., Zhao, W., Ding, N., Liu, Z., & Sun, M. (2022). PTR: Prompt tuning with rules for text classification. https://arxiv.org/abs/2105.11259
19. Ahn, M., Brohan, A., Brown, N., Chebotar, Y., Cortes, O., David, B., Finn, C., Fu, C., Gopalakrishnan, K., Hausman, K., Herzog, A., Ho, D., Hsu, J., Ibarz, J., Ichter, B., Irpan, A., Jang, E., Ruano, R. J., Jeffrey, K., Jesmonth, S., Joshi, N. J., Julian, R., Kalashnikov, D., Kuang, Y., Lee, K.-H., Levine, S., Lu, Y., Luu, L., Parada, C., Pastor, P., Quiambao, J., Rao, K., Rettinghouse, J., Reyes, D., Sermanet, P., Sievers, N., Tan, C., Toshev, A., Vanhoucke, V., Xia, F., Xiao, T., Xu, P., Xu, S., Yan, M., & Zeng, A. (2022). Do as I can, not as I say: Grounding language in robotic affordances. https://arxiv.org/abs/2204.01691
20. Spokoyny, D., Wu, C.-S., & Xiong, C. (2022). Numerical correlation in text. In D. Ferreira, M. Valentino, A. Freitas, S. Welleck & M. Schubotz (Eds.), *Proceedings of the 1st Workshop on Mathematical Natural Language Processing (MathNLP)* (pp. 33–39). Association for Computational Linguistics. https://aclanthology.org/2022.mathnlp-1.5/
21. Vaswani, A., Shazeer, N., Parmar, N., Uszkoreit, J., Jones, L., Gomez, A. N., Kaiser, L., & Polosukhin, I. (2017). Attention is all you need. https://arxiv.org/abs/1706.03762
22. Shazeer, N., & Stern, M. (2018). Adafactor: Adaptive learning rates with sublinear memory cost. https://arxiv.org/abs/1804.04235
23. Loshchilov, I., & Hutter, F. (2017). Decoupled weight decay regularization. In *International Conference on Learning Representations*. https://arxiv.org/abs/1711.05101
24. Hermans, J., Spanakis, G., & Möckel, R. (2017). Accumulated gradient normalization. https://arxiv.org/abs/1710.02368
25. Tirumala, K., Markosyan, A., Zettlemoyer, L., & Aghajanyan, A. (2022). Memorization without overfitting: Analyzing the training dynamics of large language models. https://arxiv.org/abs/2205.10770

Hierarchy-Aware Entity Embeddings for Classification in Large Taxonomies

Elena Chistova and Ivan Smirnov

1 Introduction

Hierarchical classification involves organizing data into a structured hierarchy of categories, where each category is a subset of a broader parent category. Hierarchical text classification (HTC) is extensively used to organize either elaborate textual documents or short entities. This work addresses the challenge of inferring fine-grained categories based solely on an entity's name, a particularly demanding task within hierarchical classification. Accurate classification in such scenarios is demanding, especially when important context is limited or unavailable. Effective entity encoding methods that leverage extensive taxonomies are essential for enhancing classification performance and ensuring controllability across real-world applications.

Recent advancements in deep hierarchical text classification [7, 8, 24] have significantly improved the ability to handle complex hierarchical structures. However, these methods exhibit three major limitations: (1) dependence on multiple attributes beyond the entity text itself, which is commonly adopted in e-commerce where such attributes are often available, (2) limited applicability to unbalanced data and underrepresented taxonomy branches, and (3) controllability constraints. To overcome these challenges, we propose a novel approach that incorporates information about annotated taxonomy into entity encodings. Our method focuses on encoding the textual title of an entity in a manner that reflects its position within a large and complex taxonomy, thereby facilitating accurate and efficient mapping to the relevant leaf nodes. Our proposed method offers several key advantages:

E. Chistova (✉) · I. Smirnov
FRC CSC RAS, Moscow, Russia
e-mail: chistova@isa.ru; ivs@isa.ru

M. Abbas (ed.), *Mastering Text Classification*, Signals and Communication Technology, https://doi.org/10.1007/978-3-031-93612-8_5

- By leveraging hierarchy-aware encodings, our method improves the representation of entity titles, enabling high classification performance even when only the entity name is available.
- The proposed encoding technique effectively handles large and highly imbalanced hierarchies, ensuring accurate classification of both common and rare categories.
- Our approach enhances the controllability of hierarchical classifiers, providing clearer insights into the classification process and enabling easier verification by end users.

This work builds upon our previous study on representation learning for the hierarchical classification of product titles [5], extending its application to the healthcare domain. We address the domain-specific, language-specific, and precision-critical task of classifying medical procedures in Russian into a large taxonomy with limited expert annotations of a substantial part of taxonomy leaves.

We demonstrate the practical efficacy of our approach in both healthcare and e-commerce settings. In healthcare, accurate classification of medical services and procedures is essential for automating claims processing and ensuring proper reimbursement. Our method handles non-standardized Russian medical terminologies, streamlining data alignment between healthcare providers and insurance companies. In the e-commerce, our approach enhances the organization of product and service titles, improving searchability and user experience while maintaining scalability across large product catalogs.

These applications demonstrate how hierarchy-aware entity encodings effectively integrate complex taxonomies with practical classification tasks, offering a flexible solution adaptable to various industries. The following sections provide a comprehensive overview of our methodology, the data employed, and the results obtained from implementing our approach in both healthcare and e-commerce contexts.

2 Related Work

Hierarchical classification assigns each object to a specific branch (leaf) within a taxonomy tree. There are three fundamental approaches to hierarchical classification: flat classification (object to branch), global classification, and local classification [19]. Flat classification assigns objects directly to branches without considering the hierarchical structure. In contrast, global classification employs a single model that predicts classes across the entire hierarchy by leveraging class dependencies. Local classification, meanwhile, utilizes multiple separate models, each dedicated to different nodes or levels within the hierarchy.

Previous approaches to hierarchical entity titles classification primarily focus on classifying product titles using additional descriptions. These methods include flat classifiers [20, 22], two-level pipelines [3, 6, 9, 11], multilabel classifiers [14, 25], and sequence-to-sequence branch generation [16]. Additionally, shared tasks in this domain often feature systems that explore various extensive techniques, such as model ensembling [14, 24, 25], pseudo-labeling [24], and data augmentation

[1]. Some approaches aim to optimize the classification model itself by integrating the hierarchy of classes into activation functions [24] or loss functions [7]. Other methods focus on matching an entity title with a leaf title [4, 8]. To improve entity title encoding for product classification and address domain shift issues, Brinkmann and Bizer [2] suggest additional pretraining of transformers on product offers from Common Crawl.

The latest advancements in natural language processing involve leveraging large language models (LLMs). Recent attempts to tackle the HTC task through hierarchy verbalization still rely on pretrained BERT-like language models rather than LLMs and require modifications to the model architecture followed by subsequent fine-tuning. For instance, the problem can be framed as a hierarchy-aware multilabel masked language modeling task [23], adopting a graph attention network and a zero-bounded multilabel cross-entropy loss. Similarly, Ji et al. [13] address HTC as a flat classification by verbalizing it with a hierarchy-aware decoder constraint. Although these methods show promise, they are tailored for and evaluated on elaborate texts within smaller taxonomies such as WOS, DBPedia, and RCV1-V2.

Applying LLMs to hierarchical classification of entity titles in large and unbalanced taxonomies reveals significant limitations. Firstly, treating hierarchical classification as a flat text-to-leaf task requires the LLM to memorize thousands of taxonomy branches and adhere strictly to their complex structures. This is challenging for such models due to both hallucinations and limited taxonomy understanding [21]. Enforcing the precision needed in practical applications often necessitates imposing low-level constraints that undermine the inherent generative capabilities of LLMs, rendering fine-tuned text encoders a more efficient solution. Secondly, while few-shot learning is effective in various applications, it is unsuitable for large-scale hierarchical classification due to the impracticality of exposing the model to examples across multiple thousands of taxonomy branches or the extensive resources required for fine-tuning on large labeled datasets. An alternative, albeit resource-intensive, strategy would be implementing a step-by-step hierarchical classification via LLM, where predictions at each level inform subsequent classification prompts. However, this approach is cost prohibitive due to the necessity of multiple elaborate prompts per short entity title and remains vulnerable to error accumulation across taxonomy levels. Thus, we can conclude that solving this intricate problem with LLMs remains too costly, time-consuming, and uncontrollable for real-world applications to date.

In contrast, we propose a method that trains an efficient deep encoder-based classifier and utilizes it to encode entity titles within a complex hierarchy for flat categorization. We demonstrate that this approach excels in terms of accuracy at the deepest levels of the hierarchy, as well as in simplicity and controllability.

3 Methods

In this chapter, we address the challenge of positioning incoming entity titles within a deep hierarchical taxonomy H comprising branches up to depth K. The primary objective is to develop an application that maximizes classification precision to minimize the manual workload of data operators, while simultaneously achieving higher recall than exact text matching against the existing database.

3.1 Nearest-Neighbor Matching

A straightforward approach frames hierarchical classification as a flat nearest-neighbor (k-NN) retrieval, leveraging either symbolic or pretrained embeddings. Specifically, for each incoming entity title T, we identify its k nearest neighbors from the database $\mathcal{D}$ based on cosine similarity between their respective embeddings:

$$\mathrm{NN}_k(T) = \arg\operatorname*{Top_k}_{D \in \mathcal{D}} \cos(\mathbf{e}_T, \mathbf{e}_D), \tag{1}$$

where $\mathbf{e}_T$ and $\mathbf{e}_D$ denote the embedding vectors of the input title T and a database title D, respectively.

However, the absence of domain-specific embeddings can pose significant challenges in real-world applications. This limitation often results in the retrieval of lexically similar yet semantically disparate neighbors (e.g., "neurosonography" classified under ultrasound versus "duodenography" under diagnostic radiology). To achieve satisfactory performance, it becomes necessary to empirically adjust the distance threshold between neighbors. In our healthcare data study, applying a cosine similarity threshold of 97.3% with fastText embeddings yielded a precision of 95%, with 13% of exact matches and 40% of the test data remained unclassified.

Overall, we defined several limitations of this approach:

- Domain shift affects pretrained encoders substantially, and domain adaptation requires additional resources for data collection and computation.
- With a small k, the method becomes more susceptible to outliers unavoidable in complex hierarchical annotations.
- This method does not provide informative insights into the underlying taxonomy structure.

Despite these drawbacks, the nearest-neighbor approach offers notable advantages, including controllability through distance threshold adjustments and immunity to subclass imbalance within a complex hierarchy when utilizing a small k.

In our experiments, we encode an entity title as an average of its token embeddings. We explore two types of representations: *fastText* and *BERT-based* embeddings.

3.2 Deep Hierarchical Classification

A more advanced method involves training a deep HTC model, which simultaneously predicts multiple labels corresponding to different levels of the taxonomy.

Given an input entity title T and a deep classification model C, the model predicts label distributions for each hierarchy level $l \in \{1, \ldots, K\}$. Let $B = (b_1, b_2, \ldots, b_K)$ denote a branch in the taxonomy H, where b_l represents the label at level l. The cumulative conditional probability $P(B|T)$ for branch B given the input title T is defined as

$$P(B \mid T) = \prod_{l=1}^{K} P(y_l = b_l \mid y_1, \ldots, y_{l-1}, T), \tag{2}$$

where $P(y_l = b_l \mid y_1, \ldots, y_{l-1}, T)$ is the probability of label b_l at level l conditioned on the labels of the preceding levels and the input title T. The final taxonomy branch is determined by selecting the leaf with the highest cumulative conditional probability:

$$B^* = \arg\max_{B \in \mathcal{H}} P(B \mid T), \tag{3}$$

where $\mathcal{H}$ represents the set of all possible branches in the taxonomy H.

In our experiments, we construct deep HTC models based on pretrained fastText and BERT-based embeddings.

FastText-Based Hierarchical Classification (FT_{1-K})

To predict the top-n possible nodes for levels 1 through K, we employ a one-vs-all multilabel classification approach implemented using the fastText[1] library.

BERT-Based Hierarchical Classification (BERT_{1-K})

We utilize an architecture of a deep hierarchical classifier similar to that in [7]. The output layers for each level are added on top of an encoding transformer. For a title

[1] https://fasttext.cc/

consisting of tokens $w_1, w_2, \ldots, w_z$, the representations are computed by the encoder:

$$\mathbf{e} = \text{Encoder}(w_1, w_2, \ldots, w_z) \in \mathbb{R}^{d_{LM}}. \quad (4)$$

The output for each hierarchy level l is predicted using a separate feedforward layer. The input for the output layer at level $l > 1$ is a concatenation of the text embedding $\mathbf{e}$ and the output from the previous level y_{l-1}:

$$y_l = \begin{cases} \text{FF}_l(\mathbf{e}) & \text{if } l = 1, \\ \text{FF}_l(\mathbf{e} \oplus y_{l-1}) & \text{otherwise,} \end{cases} \quad (5)$$

where $\oplus$ denotes the concatenation operation.

The probabilities of classes for a hierarchy level l are calculated by passing y_l through the softmax activation function. The class with the highest predicted probability is then selected as $\hat{y}_l$. The loss function is a weighted sum of the categorical cross-entropy loss and hierarchical loss:

$$\text{HLoss}_l = \begin{cases} 0 & \text{if } \hat{y}_l \subseteq \hat{y}_{l-1}, \\ 1 & \text{otherwise,} \end{cases}$$
$$\text{Loss} = \alpha \sum_{i=1}^{K} \text{CELoss}_i + \sum_{i=2}^{K} \beta^{i-1} \text{HLoss}_l, \quad (6)$$

where α and β are weights controlling the impact of hierarchical loss. The hyperparameter β $(0 < \beta \leq 1)$ is used to scale the hierarchical loss. The cross-entropy loss is weighted to handle class imbalance at each hierarchy level.

HTC models present multiple advantages:

1. Internal entity representations are influenced by both lexical features and their hierarchical positions, enhancing semantic understanding.
2. These models exhibit increased resilience to data noise, improving overall classification reliability.
3. Deep encoders can more or less adjust to specific domains while fine-tuning.

However, this approach also reveals certain challenges:

1. Performance can be severely affected by imbalanced class distributions within the hierarchy.
2. Achieving desired precision levels is more complex compared to nearest-neighbor methods.
3. Neural models may predict nonexistent taxonomy branches, necessitating additional constraints to ensure taxonomic validity.

3.3 Our Hybrid Approach

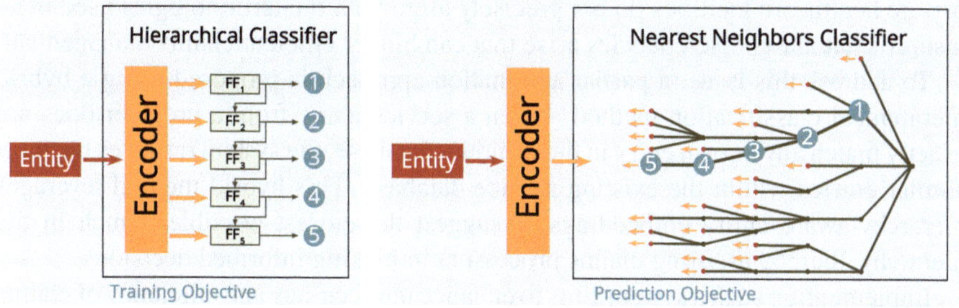

Fig. 1 Overview of our framework with BERT_{1-5} as a deep classifier. During training, the encoder is paired with outputs for hierarchical classification. The classification part of the model is fine-tuned in conjunction with the encoder to generate a sequence of subclass labels (levels 1–5). During inference, we encode the known data and input entity using only the fine-tuned encoder and attempt to find the most similar entities in a complex taxonomy. Finally, we assign the input entity to the hierarchy leaf with the most similar known entities

To balance the strengths and mitigate the limitations of the previously discussed methods, we introduce a hybrid approach that enhances an out-of-the-box text encoder through additional pretraining for hierarchical classification. The overall framework is shown in Fig. 1. In this approach, the title encoder is fine-tuned within a deep hierarchical classifier, as described in Sect. 3.2. Downstream, its function is restricted to producing entity embeddings. A nearest-neighbor classifier then processes flat (entity, leaf) pairs to predict the leaf node corresponding to the most similar entities.

The subsequent sections detail the applications of the proposed hybrid method and present the evaluation results for both the base and hybrid classifiers.

3.4 Metrics

For all experiments involving annotated data, we employed 5-fold cross validation stratified by the first-level class to maintain hierarchical balance across splits. Firstly, we calculate macro-averaged F1 for each level of the hierarchy. Since this F1 reflects performance for each level independently, we also evaluate the accuracy for flat branch assignment for each depth.

4 Service Classification in Healthcare

In healthcare, the standardization of fine-grained service classifications has not yet been universally adopted. This lack of uniformity in data, particularly when dealing with non-English languages, poses significant challenges for hospital price coordination. These challenges are especially pronounced in the *automating claims processing* required by medical insurance companies. When services provided by private healthcare facilities do not precisely align with the terminologies used in an insurer's database, discrepancies arise that can hinder efficient claims management.

To address this issue, a partial automation approach is proposed using a hybrid hierarchical classification method. When a service name from a provider does not exactly match any known entry in the insurer's database, the system retrieves the most similar entries within the existing service database. This hybrid method leverages hierarchy-aware entity embeddings to suggest the closest possible branch in the hierarchy, thereby assisting claims processors in making informed decisions.

Implementing such a system aims to enhance the accuracy and efficiency of claims processing by reducing manual labor and minimizing errors related with ambiguous service classifications. Our hybrid approach is particularly effective in determining the placement of short entity names within large, highly imbalanced hierarchies. These settings fully comply with the specified application.

4.1 Data

This chapter focuses on medical services in Russian. Data were sourced from five private healthcare facilities and one insurance company. The reference catalog of the insurance comprises 1,004 entries with a hierarchical depth of up to five levels, serving as the standard for classifying service names.

The aggregated data from the healthcare facilities include 15,815 unique entities, representing the names of services offered at each facility. The majority of these entities (95%) consist of 1 to 20 tokens, including abbreviations and acronyms, with a median length of seven tokens. Entities were manually mapped to the hierarchical catalog, as illustrated below:

Service: *Туширование гранул и боковых валиков*
(Cauterization of granules and lateral ridges)
Classification: procedures → therapeutic and/or diagnostic procedures → ENT → therapeutic procedures

The results indicated that the entities corresponded to 21 Level 1 classes, 154 Level 2 classes, 348 Level 3 classes, 188 Level 4 classes, and 36 Level 5 classes. Figure 2 illustrates the distribution of labeled examples across the first-level classes. The service names formed a highly skewed hierarchy, with 32% of the branches containing three or fewer examples.

The hierarchy includes branches with lengths ranging from one to five levels. To facilitate the prediction of shorter branches, an artificial class labeled "None" was introduced at each level starting from the second level. As the level number increases, the number of examples in the "None" class grows substantially, leading to significant class imbalance in Levels 3–5. Specifically, the frequency ratios of the "None" class to the next most represented class are 1.3:1 for Level 2, 12:1 for

Fig. 2 Distribution of labeled service names across Level 1 classes

Level 3, 23:1 for Level 4, and 485:1 for Level 5. The highly skewed nature of the hierarchy necessitated the development of a method that balances entity matching (achieving high precision and low recall) with hierarchical classification (resulting in low precision and high recall for major classes). This balance was achieved by utilizing all annotated information within the domain-specific data to produce hierarchy-aware entity embeddings.

4.2 Experimental Setup

For experiments involving healthcare entities in Russian, we utilize the following pretrained models:

fastText A continuous bag-of-words fastText model [10] was pretrained on articles within the Russian Wikipedia category Medicine. Preliminary experiments demonstrated a 2% increase in the hierarchical model's average F1 score relative to the commonly pretrained Russian fastText model. **Hierarchical model** (FT_{1-5})**:** The classifier is fine-tuned using the one-vs-all approach, with a learning rate of 1, a character n-gram range of (3, 10), and trained for 25 epochs. The top 7 predicted nodes are used to assemble the final branch post-classification.

Contextual Embeddings For contextual embeddings, we employ the pretrained transformer RuBERT[2] [15]. **Hierarchical model** (BERT_{1-5})**:** The model is fine-tuned with a learning rate of $2e-5$, a dropout rate of 0.5, a batch size of 8, $\alpha = 1$, $\beta = 0.9$, and the cross-entropy loss for each level (CELoss_i as defined in Eq. (6)), with weights adjusted based on the class distribution within each subcorpus. The top 8 predicted nodes are used to assemble the final taxonomy branch.

---

[2] `DeepPavlov/rubert-base-cased`

4.3 Experimental Results

Table 1 presents the F1 scores for node classification at each level of the healthcare services hierarchy. A pronounced skew in the hierarchy emerges starting at level 3, which is reflected by an increase in F1 scores at levels 4 and 5, following consistent declines from levels 1 through 3. The accuracy of branch assignments is detailed in Table 2.

Lvl.	KNN:FT	FT_{1-5}	KNN:FT_{1-5}	KNN:BERT	$BERT_{1-5}$	KNN:$BERT_{1-5}$
1	65.0	71.0	73.4	63.5	**83.9**	83.8
2	37.0	47.3	50.0	37.9	63.0	**65.7**
3	26.6	38.7	40.2	27.3	50.4	**55.6**
4	29.3	28.9	36.3	27.7	46.3	**50.7**
5	31.4	28.9	38.1	28.9	48.6	**52.1**
Avg.	37.7	43.0	47.6	37.1	58.4	**61.6**

Table 1 Mean F1 for healthcare service classification

Baselines We found that using averaged embeddings from an off-the-shelf language model (KNN:BERT) to retrieve the nearest services was the weakest baseline for this task. It achieved a 63.5% F1 score for classification at the first hierarchy level and an accuracy of 44.1% in predicting the correct branch. The poor performance of KNN:BERT is likely due to a known issue with transformers: the feature extraction performance of the frozen models decreases with increasing difference between pretraining and target tasks [18]. On the other hand, a simpler fastText-based encoder, which relies primarily on surface n-gram similarity, provides a more robust baseline for this task (KNN:FT). While the increase in F1 score at each level is not substantial compared to KNN:BERT (Table 1), it notably improves the ability to differentiate entities at lower hierarchy levels, reaching 51.2% accuracy in predicting the entire branch (Table 2).

	A_1	A_{1-2}	A_{1-3}	A_{1-4}	A_{1-5}
KNN:FT	75.5	62.9	53.7	51.4	51.2
FT_{1-5}	79.5	60.2	54.8	53.6	53.4
KNN:FT_{1-5}	82.3	71.9	64.8	62.9	62.6
KNN:BERT	73.6	56.2	46.8	44.3	44.1
$BERT_{1-5}$	**89.2**	75.4	68.4	66.2	66.2
KNN:$DBERT_{1-5}$	**89.2**	**79.3**	**73.9**	**72.0**	**71.7**

Table 2 Mean accuracy of the branch prediction

Deep Hierarchical Classification Results show that the trainable hierarchical classifier FT_{1-5} outperforms its corresponding baseline, especially at the first hierarchy level. The differences in node prediction performance at each level (Table 1) suggest that the overall increase in branch prediction accuracy (Table 2) is mainly driven by better distinction of the 21 major classes illustrated in Fig. 2. However, accurately classifying a large number of detailed subclasses remains challenging for the fastText classifier. Implementing hierarchical classification on top of a language model with a hierarchy-aware objective (BERT$_{1-5}$ model) led to the highest performance compared to all baseline and fastText-based approaches. The BERT$_{1-5}$ model achieved an F1 score of 83.9% for first-level nodes, outperforming FT$_{1-5}$ by 15.7% F1 and KNN:FT$_{1-5}$ by 13.0% F1. It also achieved a 66.2% accuracy in whole branch prediction. Importantly, this performance was achieved despite the language model being pretrained on elaborate common-domain texts and the dataset consisting of short medical entity titles. The dramatic performance difference compared to frozen embeddings (KNN:BERT) underscores the effectiveness of BERT-based hierarchical classification method.

Hierarchy-Aware Entity Embeddings Using internal representations of entities trained jointly with the deep hierarchical classifier enabled further improvements in classification performance. It improved node classification at the lower levels in fastText-based approaches, where KNN:FT relies heavily on n-gram matching, while FT$_{1-5}$ struggles to identify distinctive features that differentiate each node from hundreds of others, including those hierarchically related. Results for KNN:FT$_{1-5}$ show that this model successfully identified important node-related features, although it was not effective in predicting the actual branches. Its encodings proved highly effective for leaf representations, evidenced by a 9.2% increase in branch prediction accuracy. This method achieved sufficient accuracy (62.6%) with relatively modest resources, which is advantageous for practical applications. The experiments were conducted using the CPU-optimized fastText framework and did not require the development costs associated with complex architectures or fine-tuning pretrained transformers. However, fine-tuning a transformer achieved the best performance in hierarchical classification of entities. The nearest-neighbor method relying on the fine-tuned embeddings (KNN:BERT$_{1-5}$) improved the average F1 score for node classification by 3.2% and branch prediction accuracy by 5.5%, while maintaining the same performance for first-level predictions.

5 Product Classification in E-Commerce

Product classification in e-commerce presents distinct challenges compared to the specialized service classifications discussed previously. Many platforms, including various online marketplaces, allow third-party sellers to define their own subcategories, resulting in hierarchies that are not only exceptionally large—often having up to a thousand times more leaves than specialized curated taxonomies—but also prone to inaccuracies and inconsistencies.

Unlike the specific terminology used in healthcare services, product names in e-commerce typically use common language. A product name alone often does not convey its purpose, and the diversity and variability of product names are considerably greater. Additionally, product names often include refinements and descriptions aimed at improving search visibility and attracting customers. These additions make product entities more elaborate and less standardized than the precise terms used in specialized services.

In this section, we demonstrate that despite these challenges, our hybrid hierarchical classification method effectively encodes product names within large-scale, real-world hierarchies. The automated classification can be used to suggest the third-party product seller the most appropriate hierarchy branches for their product reducing manual categorization efforts but also minimizing errors associated with ambiguous taxonomy branches.

5.1 Data

In contrast to expert-annotated, closed-domain hierarchical classifications in healthcare, classifying titles within large e-commerce databases presents unique challenges. E-commerce datasets typically encompass an order of magnitude more data, and both item names and hierarchical structures are often corrupted due to entries made by third-party sellers. Previous research has shown that trainable models with millions of parameters, such as BERT, are prone to overfitting to noise and outliers in e-commerce product classification data [26]. The primary challenges in e-commerce data include frequently incomplete or misleading item descriptions and the presence of confusing or non-mutually exclusive labels within extensive taxonomies. These issues create significant obstacles for supervised learning when classifying images, descriptions, or titles. To address these challenges, we thoroughly cleaned the data for our experiments.

We use only the titles and hierarchy annotations from the Amazon review[3] [17] dataset; HTML character references in both titles and categories are decoded into Unicode. We prune subbranches leaving only nodes containing less than 13 tokens[4] and keep only subbranches appearing in the data at least four times. We selected seven major data subsets, each containing at least 90 classes annotated at the fifth level of the hierarchy. The statistics of the obtained data are described in Table 3. At each hierarchy level, we encode classes independently of previous levels. As a result, in most subsets, the number of classes decreases after level 4, with the "None" class placeholder appearing most frequently. This indicates a natural skew in the hierarchy.

[3] https://cseweb.ucsd.edu/~jmcauley/datasets/amazon_v2/

[4] We considered the longer nodes noisy because they often included non-taxonomy information, such as notes for customers (e.g., "*Please feel free to contact us if you have any special requests or questions*") or lengthy keyword-stuffed descriptions (e.g., "*My Daily Styles Stainless Steel Black Faux PU Leather Yellow Gold-Tone Latin Cross Religious Adjustable Wristband Mens Bracelet*") hardly resembling subclasses.

Part	Deduplicated Length	Unique Branches	Unique classes of each level 1	2	3	4	5
Clothing shoes and jewelry	1988301	35710	11	253	953	5263	12371
Home and kitchen	1203754	1671	13	136	539	695	292
Automotive	831549	2252	14	165	743	849	318
Sports and outdoors	809999	3414	3	43	351	1102	1281
Electronics	584136	900	16	105	290	305	133
Tools and home improvement	488042	1152	13	98	435	427	172
Industrial and scientific	132168	1796	25	301	970	496	93

Table 3 Statistics of the e-commerce corpus

5.2 Experimental Setup

For experiments involving English product names, we utilize the following pretrained models:

fastText We use a fastText model as described in [10], which is pretrained on Common Crawl and Wikipedia data. *Hierarchical model* (FT_{1-5})*:* The classifier is fine-tuned using the one-vs-all scheme, with a learning rate of 1, a character n-gram range of (3, 10), and trained for 25 epochs. The top 7 predicted nodes are selected to construct the final branch post-classification.

Contextual Embeddings As a pretrained transformer, we employ DeBERTa[5] [12]. *Hierarchical model* (DBERT_{1-5})*:* This model is fine-tuned with a learning rate of 2e-5, a dropout rate of 0.4, a batch size of 128, $\alpha = 1$, and $\beta = 0.9$. We apply cross-entropy loss for each hierarchy level (CELoss_i as defined in (6)), with weighted adjusted based on the distribution of classes within each subcorpus. The top 8 predicted nodes are used to assemble the final branch.

5.3 Experimental Results

Table 4 compares the performance of all investigated methods for hierarchical classification of products.

Baselines The macro F1 scores calculated for each level independently are illustrated in Fig. 3. The results for kNN using out-of-the-box pretrained text encoders are denoted as KNN:FT for fastText and KNN:DBERT for DeBERTa. Across all subcorpora, the flat kNN classifier utilizing fastText serves as a strong baseline. In contrast, the DeBERTa-based nearest-neighbor retriever performs poorly, consistent with results observed on the healthcare dataset.

Deep Hierarchical Classification The fastText- and DeBERTa-based classifiers are denoted as FT_{1-5} and DBERT_{1-5}, respectively. According to the results in Table 4,

[5] `microsoft/deberta_v3_base`.

	A_1	A_{1-2}	A_{1-3}	A_{1-4}	A_{1-5}	A_1	A_{1-2}	A_{1-3}	A_{1-4}	A_{1-5}
	Clothing shoes and jewelry					Home and kitchen				
KNN:FT	85.5	81.4	71.0	55.3	44.4	89.7	78.5	68.5	64.4	63.5
FT_{1-5}	84.1	76.8	64.1	45.4	31.1	91.5	80.0	68.0	62.9	60.6
KNN:FT_{1-5}	86.8	83.1	73.3	57.8	45.9	90.9	81.0	71.5	67.4	66.5
KNN:DBERT	72.7	64.3	51.3	38.9	31.8	64.5	49.6	42.2	39.8	39.8
DBERT_{1-5}	90.8	88.2	80.3	65.7	52.7	93.3	85.0	76.5	72.7	71.6
KNN:DBERT_{1-5}	**90.9**	**88.4**	**80.8**	**67.0**	**54.9**	**93.6**	**85.6**	**77.6**	**74.0**	**73.1**
	Automotive					Sports and outdoors				
KNN:FT	89.4	82.1	76.1	72.7	72.0	91.8	81.7	73.0	64.2	59.3
FT_{1-5}	88.5	80.1	72.9	67.9	66.6	90.3	78.0	67.1	56.7	50.3
KNN:FT_{1-5}	91.8	86.0	80.7	77.4	76.7	93.3	85.2	77.5	69.2	64.6
KNN:DBERT	76.9	66.9	61.3	58.7	58.3	77.0	54.5	46.0	40.8	38.0
DBERT_{1-5}	92.1	86.3	80.8	77.4	76.6	94.4	87.3	80.3	72.6	68.0
KNN:DBERT_{1-5}	**92.3**	**86.9**	**81.8**	**78.6**	**77.8**	**94.5**	**87.8**	**81.2**	**74.0**	**69.8**
	Electronics					Tools and home improv.				
KNN:FT	87.0	76.3	68.6	64.0	62.6	88.3	78.9	68.4	64.3	62.9
FT_{1-5}	87.4	74.6	64.8	58.3	56.7	89.9	79.8	69.2	63.7	62.1
KNN:FT_{1-5}	89.4	79.8	72.6	68.5	67.2	91.9	84.3	75.2	70.9	69.6
KNN:DBERT	63.9	50.3	43.6	40.4	39.6	62.0	51.7	43.9	41.7	40.8
DBERT_{1-5}	89.8	80.1	72.5	68.1	66.9	92.2	84.6	75.6	71.1	69.8
KNN:DBERT_{1-5}	**90.1**	**80.8**	**73.7**	**69.5**	**68.3**	**92.4**	**85.2**	**76.5**	**72.2**	**70.9**
	Industrial and scientific									
KNN:FT	82.0	71.8	63.6	60.6	60.2					
FT_{1-5}	85.1	74.1	64.7	60.3	59.7					
KNN:FT_{1-5}	87.9	79.1	71.4	68.2	67.8					
KNN:DBERT	56.8	49.0	44.1	42.4	42.3					
DBERT_{1-5}	88.2	78.9	70.6	67.4	67.0					
KNN:DBERT_{1-5}	**88.4**	**79.4**	**71.5**	**68.5**	**68.0**					

Table 4 Mean accuracy of the branch prediction. The datasets are listed in descending order of size (see Table 3)

the fastText-based hierarchical classifier outperforms the kNN baseline only across the smallest subcorpora, and mostly at higher levels of the hierarchy. Moreover, for larger datasets, starting with "Sports and Outdoors," multilabel fastText training becomes increasingly challenging and resource intensive. The model effectively learns the statistics of hierarchical labels, which is evidenced by applying kNN to its representations. However, collecting the taxonomy branch from a top-*n* pool of predicted labels directly proves to be impractical. DBERT_{1-5} outperforms not only the corresponding weak baseline but also the fastText-based hybrid classification KNN:FT_{1-5} on many datasets. It is also worth noting that this method handles larger datasets with extensive class sets more effectively than multilabel fastText.

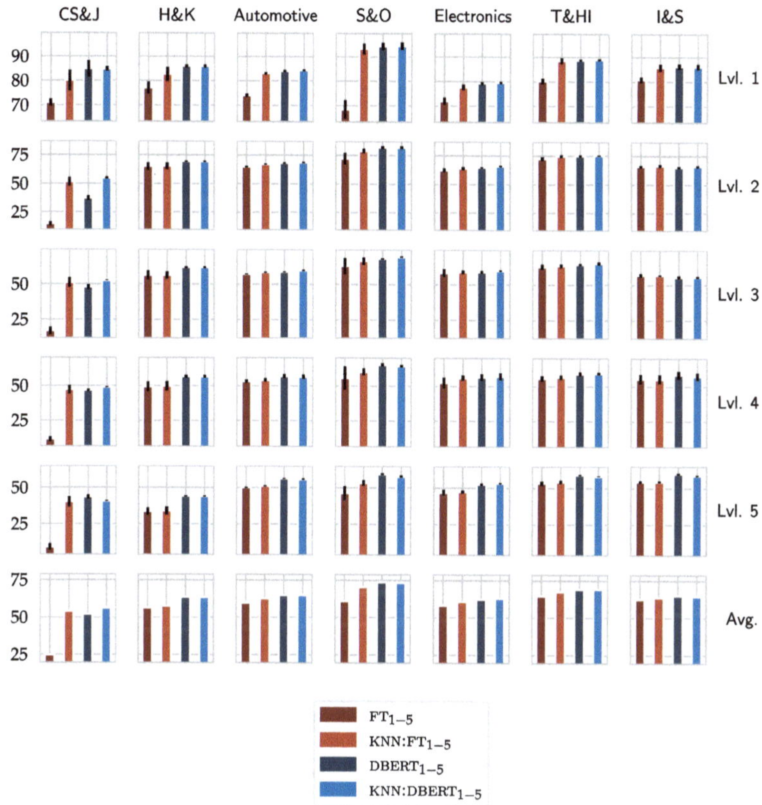

Fig. 3 Macro F1 calculated for each level independently: two types of classifiers

Hierarchy-Aware Entity Embeddings Applying kNN directly to the internal representations results in improved classification performance across all hierarchy levels for both backbones (KNN:FT_{1-5} and KNN:DBERT_{1-5}). In addition to a considerable improvement in the accuracy of full branch prediction (A_{1-5} in Table 4) while preserving or enhancing the intra-level F1 scores (Fig. 3), the purely vector-based approach is also significantly faster than assembling known branches from a pool of predicted labels for each entity in the large product hierarchy.

6 Conclusion

We propose a method for hierarchical classification of entity titles that employs representation learning to produce hierarchy-aware embeddings. These entity embeddings are applied within a nearest-neighbor flat hierarchical classification. Our baselines include pretrained encoders used as the foundational encoders in the pipeline, as well as hierarchical classifiers built with the same encoders. The proposed approach is evaluated across two languages, demonstrating its effectiveness in both healthcare and e-commerce entity classification. It successfully addresses non-standardized Russian medical terminology within a skewed, expert-annotated taxonomy of healthcare services and is applicable to detailed real-world product names in English within extensive and noisy hierarchies. Across various applications, domains, and languages, our approach outperforms corresponding deep hierarchical classification methods built on fastText and transformer-based entity encoders.

Acknowledgements The research was carried out using the infrastructure of the Shared Research Facilities «High Performance Computing and Big Data» (CKP «Informatics») of FRC CSC RAS (Moscow).

References

1. Borst, J., Korner, E., Opasjumruskit, K., & Niekler, A. (2020). Language model CNN-driven similarity matching and classification for HTML-embedded product data. In *Proceedings of the Semantic Web Challenge on Mining the Web of HTML-Embedded Product Data Co-located with the 19th International Semantic Web Conference*.
2. Brinkmann, A., & Bizer, C. (2021). Improving hierarchical product classification using domain-specific language modelling. In *Proceedings of Workshop on Knowledge Management in e-Commerce @ The Web Conference '21* (Vol. 44, pp. 14–25).
3. Cevahir, A., & Murakami, K. (2016). Large-scale multi-class and hierarchical product categorization for an E-commerce giant. In *Proceedings of COLING 2016, the 26th International Conference on Computational Linguistics: Technical Papers* (pp. 525–535). The COLING 2016 Organizing Committee.
4. Chen, H., Ma, Q., Lin, Z., & Yan, J. (2021). Hierarchy-aware label semantics matching network for hierarchical text classification. In *Proceedings of the 59th Annual Meeting of the Association for Computational Linguistics and the 11th International Joint Conference on Natural Language Processing (Volume 1: Long Papers)* (pp. 4370–4379). Association for Computational Linguistics.
5. Chistova, E. (2023). Representation learning for hierarchical classification of entity titles. In M. Abbas & A. A. Freihat (Eds.), *Proceedings of the 6th International Conference on Natural Language and Speech Processing (ICNLSP 2023)* (pp. 43–50). Association for Computational Linguistics.
6. Das, P., Xia, Y., Levine, A., Di Fabbrizio, G., & Datta, A. (2017). Web-scale language-independent cataloging of noisy product listings for E-commerce. In *Proceedings of the 15th Conference of the European Chapter of the Association for Computational Linguistics: Volume 1, Long Papers* (pp. 969–979). Association for Computational Linguistics.
7. Gao, D. (2020). Deep hierarchical classification for category prediction in E-commerce system. In *Proceedings of the 3rd Workshop on e-Commerce and NLP* (pp. 64–68). Association for Computational Linguistics.
8. Gong, S., Zhou, Z., Wang, S., Chen, F., Song, X., Cao, X., Xian, Y., & Zhu, K. (2023). Transferable and efficient: Unifying dynamic multi-domain product categorization. In *Proceedings of the 61st Annual Meeting of the Association for Computational Linguistics (Volume 5: Industry Track)* (pp. 476–486). Association for Computational Linguistics.
9. Goumy, S., & Mejri, M.-A. (2018). Ecommerce product title classification. In *Proceedings of the Workshop on eCommerce (co-located with SIGIR)*.
10. Grave, E., Bojanowski, P., Gupta, P., Joulin, A., & Mikolov, T. (2018). Learning word vectors for 157 languages. In *Proceedings of the Eleventh International Conference on Language*

Resources and Evaluation (LREC 2018), Miyazaki, Japan, May 2018. European Language Resources Association (ELRA).
11. Gupta, V., Karnick, H., Bansal, A., & Jhala, P. (2016). Product classification in E-commerce using distributional semantics. In *Proceedings of COLING 2016, the 26th International Conference on Computational Linguistics: Technical Papers* (pp. 536–546). The COLING 2016 Organizing Committee.
12. He, P., Gao, J., & Chen, W. (2021). DeBERTaV3: Improving DeBERTa using ELECTRA-style pre-training with gradient-disentangled embedding sharing.
13. Ji, K., Lian, Y., Gao, J., & Wang, B. (2023). Hierarchical verbalizer for few-shot hierarchical text classification. In *Proceedings of the 61st Annual Meeting of the Association for Computational Linguistics (Volume 1: Long Papers)* (pp. 2918–2933). Association for Computational Linguistics.
14. Jia, Y., Wang, X., Cao, H., Ru, B., & Yang, T. (2018). An empirical study of using an ensemble model in e-commerce taxonomy classification challenge. In *Proceedings of the Workshop on eCommerce (Co-Located with SIGIR)*.
15. Kuratov, Y., & Arkhipov, M. (2019). Adaptation of deep bidirectional multilingual transformers for Russian language. arXiv preprint arXiv:1905.07213.
16. Li, M. Y., Tan, L., Kok, S., & Szymanska, E. (2018). Unconstrained product categorization with sequence-to-sequence models. In *Proceedings of the Workshop on eCommerce (co-located with SIGIR)*.
17. Ni, J., Li, J., & McAuley, J. (2019). Justifying recommendations using distantly-labeled reviews and fine-grained aspects. In *Proceedings of the 2019 Conference on Empirical Methods in Natural Language Processing and the 9th International Joint Conference on Natural Language Processing (EMNLP-IJCNLP)* (pp. 188–197). Association for Computational Linguistics.
18. Peters, M. E., Ruder, S., & Smith, N. A. (2019). To tune or not to tune? Adapting pretrained representations to diverse tasks. In *Proceedings of the 4th Workshop on Representation Learning for NLP (RepL4NLP-2019)* (pp. 7–14). Association for Computational Linguistics.
19. Silla Jr, C. N., & Freitas, A. A. (2011). A survey of hierarchical classification across different application domains. *Data Mining and Knowledge Discovery, 22*, 31–72.
20. Skinner, M. (2018). Product categorization with LSTMs and balanced pooling views. In *Proceedings of the Workshop on eCommerce (co-located with SIGIR)*.
21. Sun, Y., Xin, H., Sun, K., Xu, Y. E., Yang, X., Dong, X. L., Tang, N., & Chen, L. (2024). Are large language models a good replacement of taxonomies? *Proceedings of the VLDB Endowment, 17*(11), 2919–2932.
22. Suzuki, S. D., Iseki, Y., Shiino, H., Zhang, H., Iwamoto, A., & Takahashi, F. (2018). Convolutional neural network and bidirectional LSTM based taxonomy classification using external dataset at SIGIR eCom data challenge. In *Proceedings of the Workshop on eCommerce (co-located with SIGIR)*.
23. Wang, Z., Wang, P., Liu, T., Lin, B., Cao, Y., Sui, Z., & Wang, H. (2022). HPT: Hierarchy-aware prompt tuning for hierarchical text classification. In *Proceedings of the 2022 Conference on Empirical Methods in Natural Language Processing* (pp. 3740–3751). Association for Computational Linguistics.
24. Yang, L., Shijia, E., Xu, S., & Xiang, Y. (2020). Bert with dynamic masked softmax and pseudo labeling for hierarchical product classification. In *Proceedings of the Semantic Web Challenge on Mining the Web of HTML-Embedded Product Data Co-located with the 19th International Semantic Web Conference*.
25. Yu, W., Sun, Z., Liu, H., Li, Z., & Zheng, Z. (2018). Multi-level deep learning based e-commerce product categorization. In *Proceedings of the Workshop on eCommerce (co-located with SIGIR)*.
26. Zhang, W., Lu, Y., Dubrov, B., Xu, Z., Shang, S., & Maldonado, E. (2021). Deep hierarchical product classification based on pre-trained multilingual knowledge. *IEEE Data Engineering Bulletin, 44*(2), 26–37.

Enhancing Named Entity Recognition in Modern Standard Arabic via Fine-Grained Part-of-Speech Tags

Abed Alhakim Freihat, Mourad Abbas, Tareq Alfraidi, Reyadh Sultan Alluhaibi, and Abdulmohsen Al-Thubaity

1 Introduction

Named entity recognition (NER) is the task that attempts to locate, extract, and automatically classify named entities into predefined classes or types in open-domain texts, such as newspaper articles [1]. NER has an important role in natural language processing (NLP) [2] since it is an integral component in many natural language applications [3] including question answering [4], machine translation [5], automatic summarization [6], and opinion mining [7, 8].

The challenge of NER comes from the distinct characteristics and complexities of the target language [9]. The Arabic language, which is the focus of our interest in this research, often exhibits high levels of ambiguity [10] due to its morphological complexity [11, 12] and the extensive ambiguity of Arabic words [13, 14]. For

A. A. Freihat (✉)
Mohamed bin Zayed University of Artificial Intelligence, Abu Dhabi, UAE
e-mail: abdelhakim.freihat@mbzuai.ac.ae

M. Abbas
University of Science and Technology Houari Boumediene, Algiers, Algeria
e-mail: m_abbas04@yahoo.fr

T. Alfraidi
Islamic University of Madinah, Madinah, Saudi Arabia
e-mail: t.alfraidi@iu.edu.sa

R. S. Alluhaibi
Taibah University, Madinah, Saudi Arabia
e-mail: rluhaibi@taibahu.edu.sa

A. Al-Thubaity
National Center for Artificial Intelligence, Saudi Data and Artificial Intelligence Authority, Riyadh, Saudi Arabia
e-mail: aalthubaity@sdaia.gov.sa

M. Abbas (ed.), *Mastering Text Classification*, Signals and Communication Technology, https://doi.org/10.1007/978-3-031-93612-8_6

```
<NERTAG> ::= <POSITION> "-" <CLASS>
<POSITION> ::= "B" | "I"
<CLASS> ::= "PER" | "LOC" | "ORG" | "EVENT"| ...
```

Fig. 1 CONLL tagging schema

example, the word كريم/*krym* can be an adjective (*generousness*), a proper noun (*Ka Reem*), a noun with a prefix (*like a gazelle*), or a proper noun preceded by a prefix (*like Reem*).

Named entities in Arabic do not have distinctive features [15], such as capitalization [16], that distinguish them from other words like other languages such as European languages [17]. For these and other reasons, named entity recognition in Arabic (ANER) is considered a difficult task [18, 19]. The difficulty increases especially in modern standard Arabic (MSA) texts [20] that may contain borrowed words and foreign proper names. For example, the word كريم/*krym* can also be a borrowed noun (cream). Another example is the word ماسك/*mAsk* that can be in some context a borrowed noun (mask) or a proper noun in others (Musk).

In addition to morphological complexity and word ambiguity and the agglutinative nature of Arabic words [21], it is possible for a word or a sequence of words to denote named entities that belong to different classes. For example, the word برشلونة/*br$lwnp* can be a city in some context or a football team in another. This problem is most evident when using ellipses [22, 23]. For example, the word المتوسط/*AlmtwsT* may refer to the Mediterranean Sea in some context or to a noun (the average) in another. Another example is the word الأطلسي/*Al>Tlsy* which may refer to the Pacific Ocean sometimes and refer to NATO other times.

Many approaches have been proposed to address ANER, which include rule-based approaches or a sequence labeling task in traditional machine learning (ML) approaches, deep learning (DL) approaches, and pretrained language models (PLMs).

NER approaches usually depend on annotated corpora in which the named entities are explicitly specified. The annotation schema used in such corpora is the CONLL-20035[1] annotation schema (Fig. 1):

Existing ANER corpora such as AQMAR [12], Wojood [17,24], CLEANANER-Corp [25], and CAraNER [26] primarily annotate entities without providing part-of-speech (POS) information for the surrounding text, limiting the models' ability to leverage linguistic context for disambiguation. Although prior studies have explored rule-based and machine learning approaches, many did not consider the benefits of detailed POS tagging. In addition, the annotation does not address the prefixes that may appear in the annotated entities. For example, the words ... أحمد ، وأحمد ، لأحمد ، will have the same tag (B-PER).

Ignoring the annotation of the surrounding text of the named minimizes the role of the annotated corpora on the efficiency ANER systems. In addition, postpro-

[1] http://www.cnts.ua.ac.be/conll2003/ner/annotation.txt

cessing steps will be needed to address the prefixes in discovered named entities. However, we believe that providing fine-grained POS tags aids in disambiguating the surrounding words of named entities, which, in turn, enhances the recognition and correct classification of named entities. In this chapter, we demonstrate through experiments on four tag sets applied to the same dataset how the use of POS tags improves ANER. Furthermore, we show that leveraging fine-grained POS tags and morphological features can achieve state-of-the-art performance without requiring any preprocessing, postprocessing, or feature engineering techniques. The raw/unannotated dataset used for all experiments is sourced from the ALP [27–29] dataset. Our contributions include:

1. Incorporating POS annotations into ANER datasets with varying levels of granularity
2. Demonstrating a significant improvement in F1 scores from 74.2% to 87.5% by progressively refining the tagging schemes
3. Analyzing the trade-offs between tagging complexity and model performance, providing insights into the importance of linguistic context

The remainder of this chapter is organized as follows. In Sect. 2, we review the most popular ANER approaches. In Sect. 3, we describe the datasets used in our experiments. In Sect. 4, we test machine learning (ML) models for the four tag sets. In Sect. 5, we discuss the results, and we conclude the paper in Sect. 6.

2 Related Work

ANER processing has received a growing interest from experts in the field of NLP. Thus, several tools were created to tag Arabic named entities correctly. There are, however, different approaches utilized to develop such tools. In the following, a brief review of the most popular approaches is given.

2.1 Rule Based

Rule-based approaches heavily rely on manually written grammatical rules created by linguists. Named entities are predefined to contain specific characteristics, which are contextually determined. These characteristics enable the recognition of the targeted element and its classification accordingly [18].

This method usually employs regular expressions or dictionary-based techniques to extract entities [30]. In other words, the approach developers "construct a bunch of heuristic rules and use them to parse the phrase to extract the name" [31]. This approach was to develop different approaches in ANER.

NERA [18], for instance, was built through two resources: (a) a whitelist extracted from gazetteer dictionary and (b) a finite state transduction grammar where a set of grammar rules that match ENs are described.

The authors evaluated their approach by testing it on the same data (corpora) that was used to identify the types of entities (whitelist) the approach should recognize. This procedure, as the authors reported, led to good performance for certain categories, like time, price, and measurement, since the approach was specifically trained to identify these entities from that particular dataset.

However, recognizing person and location still achieved a high average performance (e.g., 87.7% and 85.6% F1, respectively). The authors have finally determined that further efforts are necessary to enhance the approach performance. The developers of [32] used the General Architecture for Text Engineering (GATE) as a language engineering environment and the Prague Arabic Dependency TreeBank (PAMA) as language lexical resources. The results of the evaluation of the system they proposed showed better performance than those reported in some other systems such as PERA [33]. Other works that used the same method include [34], [35], [36], and [37]. Nevertheless, this approach has some limitations, such as dealing with ambiguity (e.g., the case where one expression can refer to multiple entities according to the context) and limited adaptability as new words appear which requires regular updates [30, 38].

2.2 Classical Machine Learning Algorithms

Unlike rule-based methods, which rely on manually written rules, the ML approach uses some particular algorithms to train the NER tool to make decisions on recognizing the entity names. This is supported by annotated data to train the approach how to tag the Arabic entity names successfully [39].

Several notable studies have utilized ML techniques to advance ANER. For example, [40] used the maximum entropy approach to train the ANERsys approach to successfully recognize Arabic entities. In further work and using the same approach, [41] developed an improved version of the approach that employs the conditional random fields (CRFs) method to improve the performance of the approach (they called it ANERsys 2.0) to overcome some limitations reported in the first version, especially classifying multi-word entities. The evaluation of the latest version clearly showed a better performance than the old one and other NER approaches, such as Sakhr. It indicated that the latest version scored 65.91% in the F1 test, while the old one scored 55.32% and 58.58% in Sakhr.

Continuing to improve the performance of ANER, [42] developed a new system using the conditional random fields (CRFs) method. They evaluated the system performance on ACECorpora and ANERcorp. The results showed the advantages of CRFs method as the system achieved an F1 score of 82.17%, which is better than that achieved by ANERsys 2.0. The authors, thus, claimed that this system gained state-of-the-art performance.

The system in [43] also employed CRFs in developing the Noor ANER system, specifically designed to classify named entities in Classical Arabic texts. They evaluated the system on three sub-corpora, and each corpus contains different genres: history, hadith, and jurisprudence. The results indicated that the system works better with the historical and hadith data than with jurisprudence (F1 score 99.93%, 93.86%, and 75.68%, respectively).

They concluded that the "CRF model has a very strong pattern to extract many of person names" where they are common in Arabic historical and hadith.

2.3 Deep Learning

Deep learning has the ability to automatically learn effective features from data, minimizing the need for manual design. Additionally, deep learning models can be trained efficiently using gradient descent, allowing for more complex and potentially more accurate NER systems for Arabic [44, 45]. Researchers have shown significant interest in ANER. [46], for example, proposed a deep neural network (DNN)-based model for ANER that avoids traditional handcrafted features. The model learns features directly from the data and performs well. The model was evaluated on ANERcorp dataset. The results showed that the model performed well and achieved an F measure of 88.64% outperforming conditional random fields (CRF) by a substantial margin, placing it near the state of the art for ANER tasks. [47] introduced a fully end-to-end neural network model based on BiLSTM-CRF for ANER. It uses pretrained FastText embeddings and character-based representations through a convolutional neural network (CNN).

The model achieved an F1 score of 76.65%, demonstrating its flexibility and state-of-the-art performance without external resources.

Inspired by the success of deep learning in NERs, [48] discuss a deep learning-based solution using BiLSTM-CRF and incorporating FastText embeddings and CNN-based character representations. The authors also employed a layer for sequence decoding. The model they proposed attained a state-of-the-art F1 score of 90.6%, outperforming previous systems in NERs. Finally, [49] work enhanced deep learning for ANER by incorporating additional features like morphological and syntactical data from Madamira, alongside word and character embeddings. The model proposed was evaluated on datasets extracted from ANERCorpus and AQMAR. The best performance, with an F1 score of 86.71% ANER Corpus and 79.48 on the AQMAR corpus.

Table 1 Main topics in ALP dataset

Topic	Size
News	660K
Medicine domain	457K
Wikipedia like	220K
Articles	313K
Science	226K
Literature and novels	264K
Economy and business	111K
Religious texts	22K
Misc	34K

2.4 Pretrained Language Model

NER tasks, pretrained language models (PLMs), proved a successful method that can be optimized. In essence, these PLMs are neural networks that have been trained on large data, which enables them to recognize complex entity names. Different techniques are used to employ this model, such as fine-tuned and feature-based [50].

BERT is a popular PLM used in ANER. For example, [51] have created ARABERT, an Arabic language representation system, to improve performance in various Arabic natural language understanding (NLU) tasks including NER. This model achieved 84.2% on the F1 test, demonstrating state-of-the-art performance.

Another system based on BERT is CamelBERT, which is a toolkit for different Arabic NLP tasks, including NER. It was trained in multiple Arabic varieties, namely, modern standard Arabic (MSA), dialect (DA), and Classical Arabic (CA). Accordingly, different tests were conducted to evaluate its performance.

One interesting result reported in the paper is that the approach achieved 82.4% in the F1 measure of NERs in MSA data. Performance decreased for DA and CA data, reaching 74.1% and 67.9%, respectively. Recently, [52] have developed WojoodNER based on an Arabic-CAMeLBERT-mix and trained on Wojood-Fine corpus, specifically tailored for ANER. The model scored 86.45% in the F1 measure, showing better performance than CamelBERT-mix system (80.8% F1).

3 Tag Sets

We used for our experiments ALP dataset, comprising 2.3 million tokens. The corpus was assembled from documents and text segments from a varied set of online resources in modern standard Arabic, such as medical consulting web pages, Wikipedia, news portals, novels, and social media resources. Table 1 shows the main topics and the distribution of the topics in ALP dataset.

Table 2 Named entity types in ALP dataset

Type	size
PER	33266
LOC	64580
ORG	25605
EVENT	1796
MISC	1732
AWARD	139

```
<TAG> ::= <POSTAG> | <NERTAG> | "O"
<NERTAG> ::= <PREFIX> <NERBASETAG>
<POSTAG> ::= <PREFIX> <BASEPOSTAG> <POSTFIX>
<BASENERTAG> ::= <POSITION> "-" <CLASS>
<POSITION> ::= "B" | "I"
<CLASS> ::= "PER" | "LOC" | "ORG" | "EVENT"| "MISC"
<BASEPOSTAG> ::= <NOUNTG> | <ADJECTIVE> | <VERB> | <ADVERB>
| <PREPOSITION> | <PRONTAG> | <PARTICLE>
<NOUNTG> ::= <NOUN> | <NUMBER>
<PRONTAG> ::= "PRON" | "DEM" | "REL"
<PREFIX> ::= <PREFIX> | <PROCLITIC> "+" | ""
<POSTFIX> ::= <POSTFIX> | "+" <ENCLITIC> | ""
<PROCLITIC> ::= "CONJ" | "PREP" | <PART>
<ENCLITIC> ::= <PRON> | <PART>
```

Fig. 2 ALP tagging schema

Table 2 contains the named entity types and their distribution in the ALP dataset. In the following experiments, the named entity annotation is the same and will not change in all of the following experiments.

In our experiments, we used four tagging schemes based on ALP tagging schema in Fig. 2:

1. The first schema used basic named entity annotations (e.g., B-PER, I-LOC).
2. Coarse-grained POS tags for all tokens (e.g., NOUN, VERB).
3. Morphological features added to nouns, adjectives, and verbs.
4. Fully detailed annotations with morphological features for all tags, including quantifiers and foreign words.

Alp tagging schema uses rich complex tags.

The morphology of words in this schema is represented via complex tags that correspond to their internal structure. The structure of a complex tag can be a part-of-speech or a named entity tag, possibly preceded by proclitics and/or followed by enclitics. Thus <ENCLTIC>, when present, stands for one or two clitic tags at the end of the word, and <PROCLITIC>, when present, is the combination of one to three tags out of a set of the proclitic tags at the beginning of the word.

In the previous schema, named entity tags consider prefixes in named entities. Thus, <NERTAG> includes <PREFIX> for tagging possible proclitics of named

وتقول_O لطيفة_B-PER بن_I-PER يحيي_I-PER (_O 38_O عاما_O)_O التي_O تعيش_O في_O كهف_O من_O
خمس_O غرف_O في_O إحدى_O قرى_O مطماطة_B-LOC ._O إن_O والديها_O توفيا_O وتزوجت_O أخواتها_O
فأصبحت_O وحيدة_O ._O وتضيف_O أن_O البيت_O سيضيع_O إذا_O رحلت_O عنه_O ._O

Fig. 3 An example dataset one annotation

entities. For example, the tag of أحمد is B-PER, for لأحمد is PREP+B-PER, for وأحمد is CONJ+B-PER, and CONJ+PREP+B-PER for ولأحمد.

The first tag set which contains <NERTAG> only is similar to the free available ANER datasets. The number of base tags in this tag set is 14, while the number of complex tags is 27.
An example of annotated text in dataset 1 is shown in Fig. 3.

As shown in the figure, only named entities are tagged. The tag of all other tokens is O.

3.1 *Tag Set Two*

In the second tag set, we annotate all tokens in the corpus using the following coarse-grained part of speech tags, where <BASEPOSTAG> can be one of the following:

- **NOUN**: For annotating common nouns (e.g., كتاب) including quantifiers (e.g., بعض)
- **NUMBER**: For annotating numbers including digits (e.g., ٥) and words (e.g., خمسة)
- **VERB**: For annotating verbs regardless of the verb tense, voice, or mood (e.g.,اكتب, كتبوا, or يكتبون)
- **ADJ**: For annotating adjectives including comparatives, and numerical adjectives (e.g., جميل, أجمل, or خامس)
- **ADV**: For annotating adverbs including time and location adverbs (e.g., فوق, الآن, or جدا)
- **PART**: For annotating Arabic particles Arabic (e.g., إن, لا, or قد)
- **PRON**: For annotating personal and possessive pronouns. The same tag is used to annotate separate (e.g., هو) and attached pronouns (e.g., هم in كتابهم or سألهم)
- **DEM**: For annotating demonstrative pronouns (e.g., هذا or هؤلاء)
- **REL**: For annotating relative pronouns (e.g., الذي or الذين)
- **CONJ**: For annotating conjunctions (e.g., و and أو)
- **PREP**: For annotating prepositions (e.g., في or ب in بالكتاب)

- **AB**: For annotating abbreviations (e.g., د or كغم)
- **PX**: For annotating punctuation marks (e.g., . or :)

The number of base tags in the tag set is 26, while the number of complex tags is 173.

An example of annotated text is shown in Fig. 4.

وتقول_CONJ+VERB لطيفة_B-PER بن_I-PER يحيى_I-PER (_PX 38_NUMBER عاما_NOUN)_PX التي_REL تعيش_VERB في_PREP كهف_NOUN من_PREP خمس_NUMBER غرف_NOUN في_PREP إحدى_NOUN قرى_NOUN مطماطة_B-LOC _PX إن_PART والديها_NOUN+PRON توفيا_VERB وتزوجت_CONJ+VERB أخواتها_NOUN+PRON فأصبحت_CONJ+VERB وحيدة_ADJ ._PX وتضيف_CONJ+VERB أن_PART البيت_DET+NOUN سيضيع_PART+VERB إذا_PART رحلت_VERB عنه_PREP+PRON ._PX

Fig. 4 An example dataset two annotation

3.2 Tag Set Three

The third tag set is the ALP tag set. The main differences between tag set two and this tag set are:

1. **Morphological features**: Adding morphological features for the nouns and adjectives.
2. **POS classification**: The POS of the previous tag set is now classified into more fine-grained tags. For example, the verbs are classified according to tense, voice, and mode. The nouns and adjectives are classified into several subclasses as explained below.
3. **Particle classification**: The particles are classified into 28 classes that cover most of the Arabic particles.

In the following, we present the new changes compared to tag set Two.

Nouns are classified into four classes, which are common nouns, quantifiers, numbers, and foreign nouns. In addition, common nouns are classified according to genus (feminine or masculine) and numerous (singular, dual, or plural) into seven classes. The tag set for <NOUNTAG> is shown in Fig. 5.

For example, the tag SMN is used for كتاب, SFN رحلة, DMN for كتابان, DFN for رحلتان, PFN for رحلات, PIN (broken plural) for كتب, NQ for بعض, NM for خمسة or 5, and FWN for تلفون. *Verbs* are classified according to tense (present and past), voice (present and past participle), and mood (imperative) into five classes.

The tag set for <VERB> is presented in Fig. 6.

For example, PRSV tag is used for يقول, PSTV for قال, PPRSV for يقال, PPSTV for قيل, and IMPV for قل.

```
<NOUNTG> ::= <NOUN> | <QUANTIFIER> | <NUMBER> | <FWN>
<NOUN> ::= (<NUMERUS><GENUS>|"FW"|"PI")"N"
<NUMERUS> ::= "S" | "D" | "P"
<GENUS> ::= "F" | "M"
<QUANTIFIER> ::= "NQ"
<FWN> ::= "FWN"
<NUMBER> ::= "NM"
```

Fig. 5 Noun tags in tag set three

```
<VERB> ::= (<PASSIVVOICE>?<TEMPUS>)|<VERBMODE>)"V"
<TEMPUS> ::= "PRS" | "PST"
<PASSIVVOICE>::="P"
<VERBMODE>::= "IMP"
```

Fig. 6 Verb tags in tag set three

```
<ADJECTIVE> ::= <ATTRIBUTIVEADJ> | <COMPARATIVE> | <NUMERICADJECTIVE>
<ATTRIBUTIVEADJ> ::= (<NUMERUS><GENUS>|"PI")"AJ"
<COMPARATIVE> ::= "AJCMP"
<NUMERICADJECTIVE> ::= "AJNM"
```

Fig. 7 Adjective tags in tag set three

Adjectives are classified into three classes, which are attributive, comparative, and numerical adjectives. Similarly to common nouns, attributive adjectives are further classified into five classes based on gender and number.

The tag set for <ADJECTIVE> is shown in Fig. 7.

For example, the tag SMAJ is used for قديم, SFAJ for قديمة, DMAJ for قديمان, DFAJ for قديمتان, PMAJ for قديمون, PFAJ for قديمات, PIAJ for قدماء, AJCMP for أقدم, and AJNM for خامس.

Adverbs are classified into three classes, which are modifier (of a verb or adjective, location, and time adverbs).

The tag set for <ADVERB> is

```
<ADVERB> ::= "T" | "LC" | "AV"
```

For example, the tag "AV" يوميا or جدا, the tag "LC" (location adverb) for فوق, and "T" (time adverb) for الآن.

Particles are classified into 28 classes such as interrogative, conditional, and future particles. The tag set for <PARTICLE> is presented in Fig. 8.

```
<PARTICLE> :: = "ACC" | "CERT" | "COND" | "EMPH" | "EQ" | "EXC" | "EXH" |
"EXL" | "EXP" | "FUT" | "IJ" | "IMPN" | "INC" | "INDEF" | "INL" | "INTG"
| "JUS" | "NEG" | "PURP" | "Q" | "REM" | "RET" | "SUB" | "SUP" |
"SUR" | "UW" | "VOC"
```

Fig. 8 Particle tags in tags in tag set three

وتقول_C+PRSV لطيفة_B-PER بن_I-PER يحيي_I-PER (_PX 38_NM عاما_SMN)_PX التي_REL
تعيش_PRSV في_P كهف_SMN من_P خمس_NM غرف_PIN في_P إحدى_NQ قرى_PIN مطماطة_B-
LOC _PX إن_ACC والديها_DMN+PRON توفيا_PSTV وتزوجت_C+PSTV أخواتها_PIN+PRON
فأصبحت_C+PSTV وحيدة_SFAJ ._PX وتضيف_C+PRSV أن_ACC البيت_D+SMN سيضيع_FUT+PRSV
إذا_COND رحلت_PSTV عنه_P+PRON ._PX

Fig. 9 An example dataset three annotation

For example, the tag "ACC" is used for إن (accusative particle), "NEG" for لا (negation particle), and "SUR" for قد (certainty particle). A complete explanation for <PARTICLE> can be found here.[2]

There was also a small change regarding <PREP> and <CONJ> tags in the tag sets three and four. The tag <PREP> has been changed from "PREP" to "P" and <CONJ> from "CONJ" to "C."

The number of unique base tags in the annotated third dataset is 82, while the number of complex tags is 538.

An example of annotated text is shown in Fig. 9.

3.3 Tag Set Four

The main idea in the fourth tag set is to cover the morphological features in all tags. We also add a more fine-grained tags for nouns and pronouns. The main differences between this tag set and tag set three are as follows:

1. **Morphological features**: Adding morphological features for the quantifiers, numbers, foreign names, verbs, pronouns, and numerical adjectives.
2. **Pronoun classification**: The attached pronouns are classified into personal possessive pronouns.
3. **Number classification**: The numbers are classified into Arabic and foreign numbers.

This tag set contains morphological features for all tags in <BASETAG> as follows.
Nouns: In the third dataset, numbers and quantifiers were not classified based on genus and number. In addition, only singular foreign nouns were tagged as foreign nouns. However, dual and plural foreign nouns are tagged in the fourth tag set as foreign and classified based on genus and numbers. The same holds for numbers and quantifiers.
The tag set for <NOUNTG> is displayed in Fig. 10.

[2] https://github.com/afreihat/ALP

```
<NOUNTG> ::= <NOUN> | <NUMBER>
<NOUN> ::= <COMMONNOUN> | <QUANTIFIER> | <FOREIGNNOUN>
<COMMONNOUN> ::= (<NUMERUS><GENUS>|"FW"|"PI")"N"
<QUANTIFIER> ::= (<NUMERUS><GENUS>|"FW"|"PI")"NQ"
<FOREIGNNOUN> :: = "W"(<NUMERUS><GENUS>|"FW"|"PI")"N"
<NUMBER> ::= <DIGIT> | <ARABICNUMBER> | <FOREIGNNUMBER>
<DIGIT> ::= "NM"
<ARABICNUMBER> ::= (<NUMERUS><GENUS>PI)"NM"
<FOREIGNNUMBER> ::= "W"(<NUMERUS><GENUS>|"FW"|"PI")"NM"
```

Fig. 10 Noun tags in tag set four

```
<VERB>::= <ACTIVE> | <PASSIVE> | <IMPERATIVE>
<ACTIVE> ::= <ACTIVEPRESNT> | <ACTIVEPAST>
<PASSIVE> ::= <PASSIVEPRESNT> | <PASSIVEPAST>
<ACTIVEPRESNT> ::= "PRSV"<PERSON><NUMERUS><GENUS>?
<ACTIVEPAST> ::= "PSTV"<PERSON><NUMERUS><GENUS>?
<PASSIVEPRESNT> ::= "P"<ACTIVEPRESNT>
<PASSIVEPAST> ::= "P"<ACTIVEPAST>
<IMPERATIVE> ::= "IMPV"<NUMERUS><GENUS>
<PERSON> ::= "1" | "2" | "3"
```

Fig. 11 Verb tags in tag set four

Examples for the tag <QUANTIFIER>: SMNQ for أحد, SFNQ for إحدى, DMNQ for ثلثي, DFNQ for كلتا, PMNQ for كثيرون, PFNQ for كثيرات, and PINQ for كثر
Examples for the tag <FOREIGNNOUN>: WSMN for برلمان, WSFN for دمقراطية, WDMN for تلفزيونان, WDFN for شيفرتان, and WPFN for مليشيات
Examples for the tag <ARABICNUMBER>: SMNM for ثلاث, SFNM for ثلاثة, DMNM for ألفين, DFNM for مثتان, PMNM for ثلاثون, and PINM for آلاف
Examples for the tag <FOREIGNNUMBER>: WSMNM for مليون, WDMNM for مليونان, WPFNM for مليارات, and WPINM for ملايين.

Verbs: The five verb classes in the third dataset are further classified based on person, number, and genus. Thus, present and past verb classes (passive and active) are divided into 13 classes each. Imperative verbs are classified into five classes. The tag set for <VERB> is presented in Fig. 11. For example, the tag PRSV3SM is used for يقول, PSTV1P for قلنا, PPRSV3SM for يقال. PPSTV3SM for قيل, and IMPVPM for قولوا.

Adjectives: In the third dataset, only attributive adjectives were classified according to genus and number. However, numerical adjectives are classified based on genus and number. The new tag set for <NUMERICADJECTIVE> is

```
<NUMERICADJECTIVE> :: = (<NUMERUS><GENUS>)?"AJNM"
```

```
<PRONTAG> ::= <PRON> | <DEMPRON> | <RELPRON>
<PRON> ::= <PERPRON> | <POSSPRON>
<PERPRON> ::= "PRON"<PERSON><NUMERUS><GENUS>?
<POSSPRON> ::= "P"<PERPRON>
<DEMPRON> ::= "DEM"<NUMERUS><GENUS>?
<RELPRON> ::= "REL"(<NUMERUS><GENUS>)?
```

Fig. 12 Pron tags in tag set four

For example, the tag SMAJNM is used for أول, for SFAJNM أولى, for DFAJNM اثنتين, for DMAJNM اثنان, and for PMAJNM عشرين.

Personal, Demonstrative, and Relative Pronouns. In this dataset, we distinguish between personal and possessive pronouns. Both classes are divided into subclasses based on person, number, and genus. Similarly, demonstrative and relative pronouns are divided into subclasses based on person, number, and genus. The tag set for <PRONTAG> is shown in Fig. 12. Examples for the tag <DEMPRON>: DEMSM for هذا, DEMSF for هذه, DEMDM for هذان, DEMDF for هاتان, and DEMP for هؤلاء.

Examples for the tag <RELPRON>: REL is used for ما, RELSM for الذي, RELSF for التي, RELDM for اللذان, DEMDF for اللتان, RELPM for الذين, and RELPF for اللواتي.

The number of base tags in the fourth tag set is 175, while the number of complex tags is 1712.

An example of annotated text is shown in Fig. 13.

وتقول_C+PRSV3SF لطيفة_B-PER بن_I-PER يحي_I-PER (_PX 38_NM عاما_SMN)_PX التي_RELSF
تعيش_PRSV3SF في_P كهف_SMN من_P خمس_SMNM غرف_PIN في_P إحدى_SFNQ قرى_PIN
مطماطة_B-LOC _PX إن_ACC والديها_DMN+PPRON3SF توفيا_PSTV3DM وتزوجت_C+PSTV3SF
أخواتها_PIN+PPRON3SF فأصبحت_C+PSTV3SF وحيدة_SFAJ ._PX وتضيف_C+PRSV3SF أن_ACC
البيت_D+SMN سيضيع_FUT+PRSV3SM إذا_COND رحلت_PSTV3SF عنه_P+PRON3SM ._PX

Fig. 13 An example dataset four annotation

4 Experiments

This research aims to evaluate the impact of POS tags on ANER datasets and assess how fine-grained tagging improves accuracy.

To achieve these objectives, we have conducted four experiments, where we used the four tag set schemas from the previous Section on ALP raw dataset. The NER models were trained using the OpenNLP[3] maximum entropy tagger with default hyperparameters: context window size = 3, cutoff = 3. While OpenNLP was chosen for simplicity, future work could explore transformer-based models like AraBERT or CamelBERT for potentially higher accuracy, albeit with greater computational cost.

For testing the resulting four models, we used a test dataset which contains several news articles taken from the most two famous news portals in the Arabic world, Aljazeera[4] and Alarabiya.[5]

The size of the dataset is 9K tokens and contains 875 named entities as specified in Table 3.

Table 3 Named entity types in the test dataset

Named entity type	Size
PER	296
LOC	312
ORG	233
Event and Misc	34

In the process of the results evaluation, a named entity is considered to be correctly recognized if the following three conditions hold:

1. **Correct entity**: It is a named entity.
2. **Correct type**: The type of the recognized entity is correct.
3. **Correct boundaries**: The boundaries of the recognized entity are correct.

An example of a violation for the first condition: Recognizing سيندرج as LOC in فإن ما قامت به سيندرج تحت فتوى أصدرها المفتي محمد الخلايلة.

An example of a violation for the second condition: The entity الأطلسي was rejected because it was recognized as LOC instead of ORG in أن تتخلى أوكرانيا عن محاولتها الانضمام إلى الأطلسي.

Examples of violation for the third condition: Recognizing بالمعهد الوطني لبحوث الصحة وقاريء instead of بالمعهد الوطني لبحوث الصحة in وقال غراهام كوك ، أستاذ باحث بالمعهد الوطني لبحوث الصحة وقاريء في الأمراض المعدية من جامعة إمبريال

[3] https://opennlp.apache.org/

[4] https://www.aljazeera.net

[5] https://www.alarabiya.net

كوليدج لندن ، إن نتائج الدراسة« مشجعة ». Another example is recognizing طالبان
الولايات المتحدة as one entity instead of two different entities in كيف كبدت طالبان
الولايات المتحدة خسائر تقدر بترليون دولار ؟. Another example is recognizing بي سي
instead of بي بي سي in وقال وحيد الل أماني ، المتحدث باسم حاكم إقليم ساريبول ، بي بي سي
لبي بي سي إن الهجمات بدأت مساء واستمرت إلى غاية منتصف الليل.

The test dataset was manually annotated. We used the manual annotation as a golden standard to evaluate the results of the models as explained below.

In Table 4, we show the evaluation results of the tag set one.

Table 4 Evaluation of tag set one

Entity type	Precision	Recall	F1 score
PER	61.8%	75.9%	68.1%
LOC	79.4%	81.3%	80.3%
ORG	69.5%	79.0%	73.9%
EVENT and MISC	58.8%	76.9%	66.6%
All	70.0%	78.8%	74.2%

The overall F1 score for dataset one is 74.2%. We see that LOC recognition, followed by ORG, achieved the best results, while the recognition of the other types had a poor score, especially EVENT and MISC.

In Table 5, we show the evaluation results of the tag set two.

Table 5 Evaluation of tag set two

Entity type	Precision	Recall	F1 score
PER	72.9%	83.3%	77.8%
LOC	80.7%	82.3%	81.5%
ORG	76.3%	84.3%	80.1%
EVENT and MISC	70.5%	88.8%	78.6%
All	76.5%	83.4%	79.8%

The overall F1 score for dataset two improved from 74.2% to 79.9%. The results of all entities show that just adding coarse-grained POS to the surrounding text improves the ANER results. This can be seen in all entity types except LOC which improved 1.3% only.

In Table 6, we test the effects of adding morphological features to nouns and adjectives, and the classification of verbs and articles in tag set three.

Table 6 Evaluation of tag set three

Entity type	Precision	Recall	F1 score
PER	82.7%	86.2%	84.4%
LOC	93.2%	92.0%	92.6%
ORG	78.1%	84.2%	81.0%
EVENT and MISC	73.5%	80.6%	76.9%
All	84.9%	87.7%	86.2%

The results shown in Table 6 show a significant improvement for all entity types, especially the entity type LOC which has the highest F1 score. The improvement of the achieved results is significant compared to the results of the tag set one.

Finally, Table 7 provides the results achieved by the most fine-grained tags used in dataset four.

Table 7 Evaluation of tag set four

Entity type	Precision	Recall	F1 score
PER	84.1%	87.0%	85.5%
LOC	92.9%	91.7%	92.3%
ORG	81.1%	87.5%	84.1%
EVENT and MISC	76.4%	89.6%	82.5%
All	86.1%	89%	87.5%

With F1 score 87.5%, we see that tag set four achieved the best overall result. We also see the improvement in the entity types results except for LOC which was less than tag set three. Compared to the previous results, we see that the improvement in the result is modest compared to the difference between tag set one and tag set two or tag set two and tag set three. We think that this is expected due to the large increase in the number of tags in tag set four. We believe that a better improvement requires a significant increase in the annotated data in the fourth training dataset. However, the improvement between the training datasets one and four is significant. This very significant improvement shows us the importance of using fine-grained POS tags in improving ANER.

5 Discussion

The results of the experiments confirmed that adding POS tags consistently improves performance. Dataset one achieved an F1 score of 74.2%, which increased to 79.8% with coarse-grained tags (dataset two). Introducing morphological features in dataset three further boosted the score to 86.2%, and dataset four achieved the highest F1 score of 87.5%. Despite these improvements, the modest gain between datasets three and four highlights the need for larger training datasets to fully leverage the benefits of detailed annotations.

Boundary mismatches were frequent, particularly for compound entities (e.g., 'United States of America' split into partial entities). Misclassification occurred more often in the EVENT and MISC categories due to limited training data. LOC entities consistently outperformed others, likely reflecting dataset bias.

Scaling to dialectal Arabic and larger datasets poses challenges due to linguistic variation and annotation costs. Future research should explore pretrained transformer models to address these scalability issues while maintaining computational efficiency.

The enhanced NER system can significantly improve downstream NLP tasks. For instance, machine translation can benefit from accurate entity recognition to preserve semantic integrity, while sentiment analysis can leverage precise identification of entities central to opinion mining. This makes the proposed method applicable across diverse real-world scenarios.

6 Conclusion

This study demonstrates that using fine-grained POS annotations can improve ANER performance. By systematically refining tagging schemes, we achieved a substantial improvement in F1 scores, highlighting the critical role of linguistic context. Future work will focus on scaling the approach to larger datasets, exploring dialectal Arabic, and leveraging advanced models like AraBERT to further enhance results.

Our hypothesis in this chapter was that providing fine-grained POS annotations to the surrounding text of named entities in the training data improves the accuracy of ANER. To prove our hypothesis, we have conducted four experiments. In the first experiment, we did not provide POS annotation the named entities surrounding text. In the second experiment, we annotated the surrounding text with coarse-grained POS annotations. In the third experiment, we have added morphological features to the tags of nouns and adjectives and added more fine-grained tags to verbs and particles. In the fourth dataset, we have added morphological features to all POS tags that have morphological features and used more fine-grained classifications for nouns, numbers, adjectives, and pronouns.

We have evaluated the different tag sets to test the effect of using part-of-speech tags in named ANER and how using fine-grained tags on will improve the accuracy of ANER.

The results of the experiments have proven our hypothesis, and we could jump from 74.2% F1 score in dataset one which contains no POS tags to 87.5% in dataset four which contains the fine-grained POS tags. This huge improvement in the results encourages us to add more annotated data to dataset four. We think that adding more annotated data enables us to achieve better results beyond the state-of-the-art results.

References

1. Darwish, K., Habash, N., Abbas, M., Al-Khalifa, H., Al-Natsheh, H. T., Bouamor, H., Bouzoubaa, K., Cavalli-Sforza, V., El-Beltagy, S. R., El-Hajj, W., & Jarrar, M. (2021). A panoramic survey of natural language processing in the Arab world. *Communications of the ACM, 64*, 72–81.
2. Shaalan, K. (2014). A survey of Arabic named entity recognition and classification. *Computational Linguistics, 40*, 469–510.
3. Nadeau, D., & Sekine, S. (2007). A survey of named entity recognition and classification. *Journal of Linguisticae Investigationes, 30*(1), 1–20.
4. Qwaider, M. R. H., Freihat, A. A., & Giunchiglia, F. (2017). TrentoTeam at SemEval-2017 task 3: An application of Grice maxims in ranking community question answers. In S. Bethard, M. Carpuat, M. Apidianaki, S. M. Mohammad, D. Cer & D. Jurgens (Eds.), *Proceedings of the 11th International Workshop on Semantic Evaluation (SemEval-2017)* (pp. 271–274). Association for Computational Linguistics.
5. Azab, M., Bouamor, H., Mohit, B., & Oflazer, K. (2013). Dudley North visits North London: Learning when to transliterate to Arabic. In L. Vanderwende, H. Daumé III, and K. Kirchhoff (Eds.), *Proceedings of the 2013 Conference of the North American Chapter of the Association for Computational Linguistics: Human Language Technologies.* (pp. 439–444). Association for Computational Linguistics.
6. Desai, M. R., Gachhinakatti, B., Balaganur, P., & Rathod, L. (2021). Automatic text summarization in natural language processing. In *2021 IEEE International Conference on Mobile Networks and Wireless Communications (ICMNWC)* (pp. 1–5, 2021).
7. Boudad, N., Faizi, R., Thami, R. O. H., & Chiheb, R. (2018). Sentiment analysis in Arabic: A review of the literature. *Ain Shams Engineering Journal, 9*(4), 2479–2490.
8. Bensoltane, R., Zaki, T. (2022). Aspect-based sentiment analysis: An overview in the use of Arabic language. *Artificial Intelligence Review, 56*, 2325–2363.
9. Zaghouani, W. (2012). RENAR: A rule-based Arabic named entity recognition system. *ACM Transactions on Asian Language Information Processing, 11*, 1–13.
10. Habash, N., Rambow, O. (2005). Arabic tokenization, part-of-speech tagging and morphological disambiguation in one fell swoop. In K. Knight, H. T. Ng & K. Oflazer (Eds.), *Proceedings of the 43rd Annual Meeting of the Association for Computational Linguistics (ACL'05)* (pp. 573–580), Association for Computational Linguistics.
11. Attia, M., Samih, Y., Shaalan, K., & Van Genabith, J. (2012). The Floating Arabic Dictionary: An automatic method for updating a lexical database through the detection and lemmatization of unknown words. In M. Kay & C. Boitet (Eds.), *Proceedings of COLING 2012* (pp. 83–96). The COLING 2012 Organizing Committee.
12. Mohit, B., Schneider, N., Bhowmick, R., Oflazer, K., & Smith, N. A. (2012). Recall-oriented learning of named entities in Arabic Wikipedia. In W. Daelemans (Ed.), *Proceedings of the 13th Conference of the European Chapter of the Association for Computational Linguistics* (pp. 162–173). Association for Computational Linguistics.
13. Darwish, K., Mubarak, H., & Abdelali, A. (2017). Arabic diacritization: Stats, rules, and hacks. In N. Habash, M. Diab, K. Darwish, W. El-Hajj, H. Al-Khalifa, H. Bouamor, N. Tomeh, M. El-Haj & W. Zaghouani (Eds.) *Proceedings of the Third Arabic Natural Language Processing Workshop* (pp. 9–17). Association for Computational Linguistics.
14. Pasha, A., Al-Badrashiny, M., Diab, M. T., El Kholy, A., Eskander, R., Habash, N., Pooleery, M., Rambow, O., & Roth, R. (2014). MADAMIRA: A fast, comprehensive tool for morphological analysis and disambiguation of Arabic. In N. Calzolari, K. Choukri, T. Declerck, H. Loftsson, B. Maegaard, J. Mariani, A. Moreno, J. Odijk & S. Piperidis (Eds.), *Proceedings of the Ninth International Conference on Language Resources and Evaluation (LREC'14)* (pp. 1094–1101). European Language Resources Association (ELRA).
15. Benajiba, Y., Diab, M., & Rosso, P. (2008). Arabic named entity recognition using optimized feature sets. In *Proceedings of the Conference on Empirical Methods in Natural Language Processing*, EMNLP '08, (USA) (pp. 284–293). Association for Computational Linguistics.

16. Benajiba, Y., Diab, M., & Rosso, P. (2008). Arabic named entity recognition using optimized feature sets. In M. Lapata & H. T. Ng (Eds.), *Proceedings of the 2008 Conference on Empirical Methods in Natural Language Processing* (pp. 284–293), Association for Computational Linguistics.
17. Jarrar, M., Khalilia, M., & Ghanem, S. (2022). Wojood: Nested Arabic named entity corpus and recognition using BERT. In N. Calzolari, F. Béchet, P. Blache, K. Choukri, C. Cieri, T. Declerck, S. Goggi, H. Isahara, B. Maegaard, J. Mariani, H. Mazo, J. Odijk & S. Piperidis (Eds.), *Proceedings of the Thirteenth Language Resources and Evaluation Conference* (pp. 3626–3636). European Language Resources Association.
18. Shaalan, K., & Raza, H. (2009). NERA: Named entity recognition for Arabic. *Journal of the Association for Information Science and Technology, 60*, 1652–1663.
19. Balla, H. A. M., & Delany, S. J. (2020). Exploration of approaches to Arabic named entity recognition. In *CLEOPATRA@ESWC*.
20. Salah, R. E., Qadri binti Zakaria, L. (2017). A comparative review of machine learning for Arabic named entity recognition. *International Journal on Advanced Science, Engineering and Information Technology, 7* 511–518.
21. Benajiba, Y., & Rosso, P. (2008). Arabic named entity recognition using conditional random fields.
22. Chae, J., Jung, Y., Lee, T., Jung, S., Huh, C., Kim, G., Kim, H., & Oh, H. (2014). Identifying non-elliptical entity mentions in a coordinated np with ellipses. *Journal of Biomedical Informatics, 47*, 139–152.
23. Freihat, A. A., Dutta, B., & Giunchiglia, F. (2015). Compound noun polysemy and sense enumeration in WordNet.
24. Jarrar, M., Abdul-Mageed, M., Khalilia, M., Talafha, B., Elmadany, A., Hamad, N., & Omar, A. (2023). WojoodNER 2023: The first Arabic named entity recognition shared task. In H. Sawaf, S. El-Beltagy, W. Zaghouani, W. Magdy, A. Abdelali, N. Tomeh, I. Abu Farha, N. Habash, S. Khalifa, A. Keleg, H. Haddad, I. Zitouni, K. Mrini & R. Almatham (Eds.), *Proceedings of ArabicNLP 2023* (pp. 748–758). Association for Computational Linguistics.
25. AlDuwais, M., Al-Khalifa, H., & AlSalman, A. (2024). CLEANANERCorp: Identifying and correcting incorrect labels in the ANERcorp dataset. In H. Al-Khalifa, K. Darwish, H. Mubarak, M. Ali & T. Elsayed (Eds.), *Proceedings of the 6th Workshop on Open-Source Arabic Corpora and Processing Tools (OSACT) with Shared Tasks on Arabic LLMs Hallucination and Dialect to MSA Machine Translation @ LREC-COLING 2024* (pp. 13–19). ELRA and ICCL.
26. Al-Thubaity, A., Alkhereyf, S., Alzahrani, W., & Bahanshal, A. (2022). CAraNER: The COVID-19 Arabic named entity corpus. In H. Bouamor, H. Al-Khalifa, K. Darwish, O. Rambow, F. Bougares, A. Abdelali, N. Tomeh, S. Khalifa & W. Zaghouani (Eds.), *Proceedings of the Seventh Arabic Natural Language Processing Workshop (WANLP)* (pp. 1–10). Association for Computational Linguistics.
27. Freihat, A. A., Bella, G., Abbas, M., Mubarak, H., & Giunchiglia, F. (2023). *ALP: An Arabic linguistic pipeline* (pp. 67–99). Springer.
28. Freihat, A. A., Bella, G., Mubarak, H., & Giunchiglia, F. (2018). A single-model approach for Arabic segmentation, POS tagging, and named entity recognition. In *2018 2nd International Conference on Natural Language and Speech Processing (ICNLSP)* (pp. 1–8).
29. Freihat, A. A., Abbas, M., Bella, G., & Giunchiglia, F. (2018). Towards an optimal solution to lemmatization in Arabic. *Procedia Computer Science, 142*, 132–140. Arabic Computational Linguistics.
30. Raza, S., Garg, M., Reji, D. J., Bashir, S. R., & Ding, C. (2024). Nbias: A natural language processing framework for bias identification in text. *Expert Systems with Applications, 237*, 121542.
31. Abuleil, S. () "Extracting names from Arabic text for question-answering systems," in *Coupling Approaches, Coupling Media and Coupling Languages for Information Retrieval*, RIAO '04 (pp. 638–647). Le Centre De Hautes Etudes Internationales D'informatique Documentaire.
32. Elsebai, A., Meziane, F., & Belkredim, F. Z. (2009). A rule based persons names Arabic extraction system.

33. Shaalan, K., & Raza, H. (2007). Person name entity recognition for Arabic. In V. Cavalli-Sforza & I. Zitouni (Eds.), *Proceedings of the 2007 Workshop on Computational Approaches to Semitic Languages: Common Issues and Resources* (pp. 17–24). Association for Computational Linguistics.
34. Maloney, J., & Niv, M. (1998). TAGARAB: A fast, accurate Arabic name recognizer using high-precision morphological analysis. In *Computational Approaches to Semitic Languages.*
35. Traboulsi, H. (2009). Arabic named entity extraction: A local grammar-based approach. In *2009 International Multiconference on Computer Science and Information Technology* (pp. 139–143).
36. Harrag, F., El-Qawasmeh, E., & Salman Al-Salman, A. M. (2011). Extracting named entities from prophetic narration texts (hadith). In J. M. Zain, W. M. B. Wan Mohd & E. El-Qawasmeh (Eds.), *Software Engineering and Computer Systems* (pp. 289–297). Springer.
37. Abdallah, S., Shaalan, K., & Shoaib, M. (2012). Integrating rule-based system with classification for Arabic named entity recognition. In A. Gelbukh (Ed.), *Computational Linguistics and Intelligent Text Processing* (pp. 311–322). Springer.
38. Salah, R., Zakaria, L. Q. (2017). Arabic rule-based named entity recognition systems progress and challenges. *International Journal on Advanced Science, Engineering and Information Technology, 7*, 815–821.
39. Qu, X., Gu, Y., Xia, Q., Li, Z., Wang, Z., & Huai, B. (2023). A survey on Arabic named entity recognition: Past, recent advances, and future trends. *IEEE Transactions on Knowledge and Data Engineering, 36*, 943–959.
40. Benajiba, Y., Rosso, P., & Benedí, J.-M. (2009). Anersys: An Arabic named entity recognition system based on maximum entropy. In *Conference on Intelligent Text Processing and Computational Linguistics.*
41. Benajiba, Y., & Rosso, P. (2007). Anersys 2.0: Conquering the ner task for the Arabic language by combining the maximum entropy with pos-tag information. In *Indian International Conference on Artificial Intelligence.*
42. Benajiba, P. R. Y., & Diab, M. (2008). Arabic named entity recognition: An SVM-based approach. In *Proceedings of 2008 Arab International Conference on Information Technology (ACIT)* (pp. 16–18).
43. Asgari-Bidhendi, M., Minaei, B., & Jozi, H. (2012). Extracting person names from ancient Islamic Arabic texts. In *Proceedings of Language Resources and Evaluation for Religious Texts (LRE-Rel) Workshop Programme, (LREC 2012).*
44. Helwe, C., Elbassuoni, S. (2019). Arabic named entity recognition via deep co-learning. *Artificial Intelligence Review, 52*, 197–215.
45. Khan, W., Daud, A., Khan, K., Muhammad, S., & Haq, R. (2023). Exploring the frontiers of deep learning and natural language processing: A comprehensive overview of key challenges and emerging trends. *Natural Language Processing Journal, 4*, 100026.
46. Gridach, M. (2016). Deep learning approach for Arabic named entity recognition. In A. F. Gelbukh, (Ed.), *Computational Linguistics and Intelligent Text Processing - 17th International Conference, CICLing 2016, Konya, Turkey, April 3–9, 2016, Revised Selected Papers, Part I*, vol. 9623 of Lecture Notes in Computer Science (pp. 439–451). Springer.
47. Awad, D., Sabty, C., Elmahdy, M., & Abdennadher, S. (2018). Arabic name entity recognition using deep learning (pp. 105–116).
48. Bazi, I. E., & Laachfoubi, N. (2019). Arabic named entity recognition using deep learning approach. *International Journal of Electrical and Computer Engineering (IJECE), 9*(3), 2025–2032.
49. Hatab, A. L., Sabty, C., & Abdennadher, S. (2022). Enhancing deep learning with embedded features for Arabic named entity recognition. In N. Calzolari, F. Béchet, P. Blache, K. Choukri, C. Cieri, T. Declerck, S. Goggi, H. Isahara, B. Maegaard, J. Mariani, H. Mazo, J. Odijk & S. Piperidis (Eds.), *Proceedings of the Thirteenth Language Resources and Evaluation Conference* (pp. 4904–4912). European Language Resources Association.
50. Devlin, J., Chang, M.-W., Lee, K., & Toutanova, K. (2019). Bert: Pre-training of deep bidirectional transformers for language understanding. In *North American Chapter of the Association for Computational Linguistics.*

51. Antoun, W., Baly, F., & Hajj, H. (2020). AraBERT: Transformer-based model for Arabic language understanding. In H. Al-Khalifa, W. Magdy, K. Darwish, T. Elsayed & H. Mubarak (Eds.), *Proceedings of the 4th Workshop on Open-Source Arabic Corpora and Processing Tools, with a Shared Task on Offensive Language Detection* (pp. 9–15). European Language Resource Association.
52. Alshammari, N. (2024). Bangor University at WojoodNER 2024: Advancing Arabic named entity recognition with CAMeLBERT-mix. In N. Habash, H. Bouamor, R. Eskander, N. Tomeh, I. Abu Farha, A. Abdelali, S. Touileb, I. Hamed, Y. Onaizan, B. Alhafni, W. Antoun, S. Khalifa, H. Haddad, I. Zitouni, B. AlKhamissi, R. Almatham & K. Mrini (Eds.), *Proceedings of The Second Arabic Natural Language Processing Conference* (pp. 880–884). Association for Computational Linguistics.

Index

M. Abbas (ed.), *Mastering Text Classification*, Signals and Communication Technology, https://doi.org/10.1007/978-3-031-93612-8

R

S

T

Z

MIX
Papier aus verantwortungsvollen Quellen
Paper from responsible sources
FSC® C105338
FSC
www.fsc.org

If you have any concerns about our products,
you can contact us on
ProductSafety@springernature.com

In case Publisher is established outside the EU,
the EU authorized representative is:
Springer Nature Customer Service Center GmbH
Europaplatz 3, 69115 Heidelberg, Germany

Printed by Libri Plureos GmbH
in Hamburg, Germany